COMPUTATIONAL WAVE DYNAMICS

ADVANCED SERIES ON OCEAN ENGINEERING

Series Editor-in-Chief
Philip L- F Liu (*Cornell University*)

Published

Vol. 26 Hydrodynamics Around Cylindrical Structures (Revised Edition)
by B. Mutlu Sumer and Jørgen Fredsøe (Technical Univ. of Denmark, Denmark)

Vol. 27 Nonlinear Waves and Offshore Structures
by Cheung Hun Kim (Texas A&M Univ., USA)

Vol. 28 Coastal Processes: Concepts in Coastal Engineering and Their Applications to Multifarious Environments
by Tomoya Shibayama (Yokohama National Univ., Japan)

Vol. 29 Coastal and Estuarine Processes
by Peter Nielsen (The Univ. of Queensland, Australia)

Vol. 30 Introduction to Coastal Engineering and Management (2nd Edition)
by J. William Kamphuis (Queen's Univ., Canada)

Vol. 31 Japan's Beach Erosion: Reality and Future Measures
by Takaaki Uda (Public Works Research Center, Japan)

Vol. 32 Tsunami: To Survive from Tsunami
by Susumu Murata (Coastal Development Inst. of Technology, Japan),
Fumihiko Imamura (Tohoku Univ., Japan),
Kazumasa Katoh (Musashi Inst. of Technology, Japan),
Yoshiaki Kawata (Kyoto Univ., Japan),
Shigeo Takahashi (Port and Airport Research Inst., Japan) and
Tomotsuka Takayama (Kyoto Univ., Japan)

Vol. 33 Random Seas and Design of Maritime Structures, 3rd Edition
by Yoshimi Goda (Yokohama National University, Japan)

Vol. 34 Coastal Dynamics
by Willem T. Bakker (Delft Hydraulics, Netherlands)

Vol. 35 Dynamics of Water Waves: Selected Papers of Michael Longuet-Higgins Volumes 1–3
edited by S. G. Sajjadi (Embry-Riddle Aeronautical University, USA)

Vol. 36 Ocean Surface Waves: Their Physics and Prediction (Second Edition)
by Stanisław R. Massel (Institute of Oceanology of the Polish Academy of Sciences, Sopot, Poland)

Vol. 37 Computational Wave Dynamics
by Hitoshi Gotoh (Kyoto University, Japan),
Akio Okayasu (Tokyo University of Marine Science and Technology, Japan)
and Yasunori Watanabe (Hokkaido University, Japan)

*For the complete list of titles in this series, please write to the Publisher.

Advanced Series on Ocean Engineering — Volume 37

COMPUTATIONAL WAVE DYNAMICS

Hitoshi Gotoh
Kyoto University, Japan

Akio Okayasu
Tokyo University of Marine Science and Technology, Japan

Yasunori Watanabe
Hokkaido University, Japan

World Scientific

NEW JERSEY • LONDON • SINGAPORE • BEIJING • SHANGHAI • HONG KONG • TAIPEI • CHENNAI

Published by

World Scientific Publishing Co. Pte. Ltd.

5 Toh Tuck Link, Singapore 596224

USA office: 27 Warren Street, Suite 401-402, Hackensack, NJ 07601

UK office: 57 Shelton Street, Covent Garden, London WC2H 9HE

Library of Congress Cataloging-in-Publication Data

Gotoh, Hitoshi.

Computational wave dynamics / Hitoshi Gotoh (Kyoto University, Japan), Akio Okayasu (Tokyo University of Marine Science and Technology, Japan), Yasunori Watanabe (Hokkaido University, Japan).

pages cm. -- (Advanced series on ocean engineering; v. 37)

ISBN 978-9814449700 (alk. paper)

1. Wave equation--Numerical solutions. 2. Hydraulic engineering--Data processing. 3. Fluid dynamics--Data processing. I. Okayasu, Akio. II. Watanabe, Yasunori. III. Title.

TC157.8.G68 2013

551.46'30151--dc23

2013000726

British Library Cataloguing-in-Publication Data

A catalogue record for this book is available from the British Library.

In-house Editor: Amanda Yun

Printed in Singapore by B & Jo Enterprise Pte Ltd

Preface

Hydraulic engineering deals with problems involving water flows and the nature of those flows. Independent fluid motions in the atmosphere and at the water surface are mechanically coupled through momentum transfer via a dynamic force balance at the contact surface of these air-water phases, called an interface. Because the density values of air and water differ by three orders of magnitude, any mechanical contributions (including momentum transfer) across the interface is often ignored; that is, no air phase is assumed beyond the moving edge of the water region (which is called a free-surface). While analysis of the former liquid-gas two-phase flow, or interfacial flow, requires explicit contributions between both phases, the latter liquid single-phase flow, a free-surface flow, is analyzed with the mathematical framework of a simple boundary value problem of partial differential equations consisting of the equation of continuity and the Navier-Stokes equation.

Water wave mechanics is known to be based on the theoretical approach of this moving boundary problem. Many kinds of theoretical wave equations, such as the well-known mild-slope, the long-wave, and Boussinesq type equations, have been used to computationally predict wave transformation in arbitrary coastal areas. Recently, rapid developments in computer technology can reduce expensive computations for primitive equations systems, which has led to new research and findings for complex free-surface flows. A computer simulation is expected to be used for numerical experiments of coastal waves as an alternative to expensive physical experiments — potentially contributing to practical designs of coastal structures similar to the use of a numerical wind-tunnel to design an airplane or a spaceship. An application of the numerical wave flume for design support has already been undertaken; CADMAS-SURF (SUper Roller Flume for Computer Aided Design of MAritime Structure) is the numerical wave flume developed in Japan and interpreted in Chapter 4. Further developments in numerical solvers with polished computing techniques for free-surface flows are also summarized in this book.

With regard to computational design of coastal structures: researchers, engineers, and students must have easy access to the variety of methodologies for the analysis of free-surface flows to understand both advantages and shortcomings of each method, and to make appropriate choices between the methods for their practical subjects. To this end, it is essential to provide the opportunity to learn the basics of these techniques systematically.

The common and essential elements behind the numerical analysis of fluids, such as the governing equations of fluid flow, turbulence models, and fundamental schemes to solve the differential equation systems are explained in Chapters 1, 2, and 3. In these chapters, the mathematical and physical aspects of the numerical wave dynamic models are based on surface capturing methods and particle methods; these aspects are comprehensively described with some visualized illustrations of computed results. In Chapter 4, the Volume of Fluid (VOF) method is explained in detail along with an illustration of the performance of the CADMAS-SURF code. In Chapter 5, theoretical aspects of the Constrained Interpolation Profile (CIP) method are explained, as well as some models using free-surface and interface dynamics. In Chapter 6, theoretical aspects of the particle method are explained with integrated descriptions of the Smoothed Particle Hydrodynamics (SPH) method and the Moving Particle Semi-implicit (MPS) method. The turbulence model for the particle method, as well as the up-to-date accurate particle method is also explained. In Chapter 7, the Distinct Element Method (DEM) is introduced to model granular materials, and rigid-body motions driven by the flows around them. Chapter 8 describes an Euler-Lagrange hybrid method to improve the resolution of interface tracking. Finally, in Chapter 9, perspectives on future coastal problems, sand transport, design of coastal structures, and coastal disasters, based on the numerical wave flume are outlined.

In this book, experts of each method provide comprehensive descriptions of the latest numerical technologies. This book is designed with step-by-step instructions to enable young researchers and students to effectively prepare and review the relevant topics in the areas of computational wave dynamics and computational fluid mechanics. The authors also believe that the contents of this book, including scientific findings and engineering applications based on up-to-date computations, will interest young researchers and students. The list of references at the end of each chapter can provide additional help for further reading on subjects of interest.

The authors show their gratitude to Ms. Katherine Cox for her help editing the manuscript. The typesetting in TEX has been done by a graduate student, Mr. Yasuo Niida, from Hokkaido University. Finally the authors would like to thank the members of the Coastal Engineering Committee of the Japan Society of Civil Engineers for their considerable support and development in computational research in Coastal Engineering.

Hitoshi Gotoh, Akio Okayasu and Yasunori Watanabe
Spring, 2013

List of Editors and Contributors

Hitoshi Gotoh (editor)
Kyoto University, Graduate School of Engineering
Katsura Campus, Nishikyo-ku, Kyoto, 615-8540, Japan
gotoh@particle.kuciv.kyoto-u.ac.jp

Akio Okayasu (editor)
Tokyo University of Marine Science and Technology
4-5-7, Konan, Minato-ku, Tokyo, 108-8477, Japan
okayasu@kaiyodai.ac.jp

Yasunori Watanabe (editor)
Hokkaido University, School of Engineering
North 13 West 8, Sapporo 060-8628, Japan
yasunori@eng.hokudai.ac.jp

Nobuhito Mori
Kyoto University, Disaster Prevention Research Institute
Uji, Kyoto, 611-0011, Japan
mori.nobuhito.8a@kyoto-u.ac.jp

Koji Kawasaki
Nagoya University, Graduate School of Engineering
Furo-cho, Chikusa-ku, Nagoya 464-8603, Japan
kawasaki@nagoya-u.jp

Hidemi Mutsuda
Hiroshima University, Faculty of Engineering
1-4-1 Kagamiyama, Higashi-Hiroshima-shi, Hiroshima, 739-8527, Japan
mutsuda@naoe.hiroshima-u.ac.jp

Eiji Harada
Kyoto University, Graduate School of Engineering
Katsura Campus, Nishikyo-ku, Kyoto, 615-8540, Japan
harada@particle.kuciv.kyoto-u.ac.jp

Takayuki Suzuki
Yokohama National University, Faculty of Urban Innovation
79-1 Tokiwadai, Hodogaya-ku, Yokohama 240-8501, Japan
suzuki-t@ynu.ac.jp

Contents

Chapter 1

Governing Equations

Hitoshi Gotoh and Akio Okayasu

Fundamental theories governing fluid flows and water waves, used in the computational wave dynamics, are interpreted in this chapter. Mathematical approximations and models for the governing equations to be solved for describing coastal waves are also introduced in §1.3.

1.1. Governing Equations for Viscous Fluids

The governing equations for fluid flow describe the conservation laws of mass and momentum. In this section, these conservation laws are described. Furthermore, the constitution equation for fluids is detailed, an equation that is necessary for closure of the equation system, and the Navier-Stokes equation are derived, and the governing equation of a viscous fluid is explained. For additional resources that discuss the topics in this chapter, some renown books on hydrodynamics include Lamb[6] and Landau and Lifshitz.[7]

1.1.1. *Equation of continuity*

A basic concept with the continuity equation involves a budget of mass, taking into account a control volume V in a flow field (see Fig. 1.1). Because the time change of the mass in the control volume is equal to the summation of outflow flux $\boldsymbol{n}\cdot(\rho\boldsymbol{u})$ through the surface of the control volume, the budget of mass in the control volume can be written as follows:

$$\iiint_V \frac{\partial \rho}{\partial t} dV = -\iint_s \boldsymbol{n}\cdot(\rho\boldsymbol{u})dS \tag{1.1}$$

where ρ is density (mass per unit volume), S is the surface of the control volume, $\boldsymbol{u}$ is the velocity vector, and $\boldsymbol{n}$ is the unit normal vector on the surface of the control volume directed toward the outside. The right side

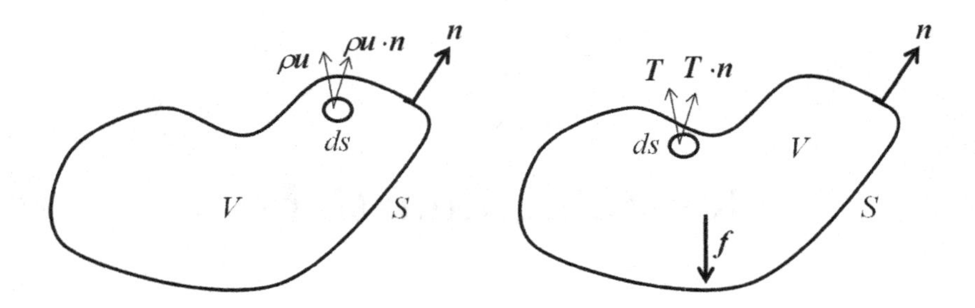

Fig. 1.1. Schematic illustrations of mass conservation (left) and momentum conservation (right) in control volume V with a surface area S.

of Eq. (1.1) is transferred to a volumetric integral of divergence of $\rho\boldsymbol{u}$ by the Gauss' theorem,

$$\iiint_V \left\{ \frac{\partial \rho}{\partial t} + \boldsymbol{\nabla} \cdot (\rho \boldsymbol{u}) \right\} dV = 0 \tag{1.2}$$

in which $\boldsymbol{\nabla}$ is defined as $\boldsymbol{\nabla} = (\partial/\partial x, \partial/\partial y, \partial/\partial z)$. Because this relation is valid for any control volume, the integrand is equal to zero.

$$\frac{\partial \rho}{\partial t} + \boldsymbol{\nabla} \cdot (\rho \boldsymbol{u}) = 0 \tag{1.3}$$

Eq. (1.3) is the differential form of the mass conservation law and is called the equation of continuity. Because ρ is constant in an incompressible fluid, the equation of continuity is simply written as follows:

$$\boldsymbol{\nabla} \cdot \boldsymbol{u} = 0 \tag{1.4}$$

1.1.2. *Momentum equation*

Focusing now on a budget of momentum, for the control volume V the momentum conservation law is written as

$$\iiint_V \frac{\partial \rho \boldsymbol{u}}{\partial t} dV = -\iint_s \boldsymbol{n} \cdot (\rho \boldsymbol{u} \otimes \boldsymbol{u} - \boldsymbol{T}) ds + \iiint_V \rho \boldsymbol{f} dV \tag{1.5}$$

where $\boldsymbol{T}$ is the stress tensor with 3×3 components, $\boldsymbol{T} = (T_{xx}\ T_{xy}\ T_{xz};\ T_{yx}\ T_{yy}\ T_{yz};\ T_{zx}\ T_{zy}\ T_{zz})$, and $\boldsymbol{f}$ is the body force (see Fig. 1.1). Tensor product $\rho\boldsymbol{u} \otimes \boldsymbol{u}$ refers to the outflow of momentum $\rho\boldsymbol{u}$ with velocity $\boldsymbol{u}$ from the control volume. Additionally, the stress tensor $\boldsymbol{T}$ incorporates any momentum exchange due to fluid stress acting on the surface of the control volume. By applying Gauss' theorem again,

$$\iiint_V \left\{ \frac{\partial \rho \boldsymbol{u}}{\partial t} + \boldsymbol{\nabla} \cdot (\rho \boldsymbol{u} \otimes \boldsymbol{u} - \boldsymbol{T}) - \rho \boldsymbol{f} \right\} dV = 0 \tag{1.6}$$

then, the differential form of the momentum equation is identically derived.

$$\frac{\partial \rho \boldsymbol{u}}{\partial t} + \boldsymbol{\nabla} \cdot (\rho \boldsymbol{u} \otimes \boldsymbol{u} - \boldsymbol{T}) - \rho \boldsymbol{f} = 0 \tag{1.7}$$

1.1.3. *Constitution equation and the Navier-Stokes equation*

To find a solution to these conservation laws, a complete equation system is necessary to achieve closure. In other words, the number of equations must coincide with the number of variables to solve the equation system. For closure, the stress tensor $\boldsymbol{T}$ should be described as a function of ρ and $\boldsymbol{u}$. This description is called a constitution equation.

In a Newtonian fluid, the stress tensor $\boldsymbol{T}$ is given by

$$\boldsymbol{T} = (-p + \lambda \boldsymbol{\nabla} \cdot \boldsymbol{u})I + 2\mu \boldsymbol{S} \tag{1.8}$$

where $\boldsymbol{I}$ is the unit tensor, p is pressure, μ and λ are physical constants, and $\boldsymbol{S}$ is the strain rate tensor defined as

$$\boldsymbol{S} = \frac{(\boldsymbol{\nabla u})^T + \boldsymbol{\nabla u}}{2} \tag{1.9}$$

The diagonal component T_{xx} of the stress tensor $\boldsymbol{T}$ is written as

$$T_{xx} = (-p + \lambda \boldsymbol{\nabla} \cdot \boldsymbol{u}) + 2\mu \frac{\partial u}{\partial x} \tag{1.10}$$

Then the average of the three diagonal components T_{xx}, T_{yy} and T_{zz} is

$$\frac{T_{xx} + T_{yy} + T_{zz}}{3} = -p + \left(\lambda + \frac{2}{3}\mu \right) \boldsymbol{\nabla} \cdot \boldsymbol{u} \tag{1.11}$$

The coefficient of the divergence of velocity $\boldsymbol{u}$ in this equation, $\lambda + (2/3)\mu$, is called the volumetric coefficient of viscosity. For cases where the media is water or air, this coefficient equals zero, $\lambda + (2/3)\mu = 0$. This is called the Stokes hypothesis, and the stress tensor $\boldsymbol{T}$ can now be written as

$$\boldsymbol{T} = -p\boldsymbol{I} + 2\mu \left\{ \boldsymbol{S} - \frac{1}{3}(\boldsymbol{\nabla} \cdot \boldsymbol{u})\boldsymbol{I} \right\} \tag{1.12}$$

For an incompressible fluid, by taking $\boldsymbol{\nabla} \cdot \boldsymbol{u} = 0$ into account, Eq. (1.12) is written as

$$\boldsymbol{T} = -p\boldsymbol{I} + 2\mu \boldsymbol{S} \tag{1.13}$$

Here the physical constant μ is called the coefficient of viscosity. By substituting Eq. (1.13) into Eq. (1.7)

$$\frac{\partial \rho \boldsymbol{u}}{\partial t} + \boldsymbol{\nabla} \cdot (\rho \boldsymbol{u} \otimes \boldsymbol{u}) = \boldsymbol{\nabla} \cdot (-p\boldsymbol{I} + 2\mu \boldsymbol{S}) + \rho \boldsymbol{f} \tag{1.14}$$

is derived. Here the left side of Eq. (1.14) is rewritten as follows:

$$\begin{aligned} \frac{\partial \rho \boldsymbol{u}}{\partial t} + \boldsymbol{\nabla} \cdot (\rho \boldsymbol{u} \otimes \boldsymbol{u}) &= \frac{\partial \rho}{\partial t}\boldsymbol{u} + \rho \frac{\partial \boldsymbol{u}}{\partial t} + \boldsymbol{u}\boldsymbol{\nabla} \cdot (\rho \boldsymbol{u}) + \rho \boldsymbol{u} \cdot \boldsymbol{\nabla} \boldsymbol{u} \\ &= \boldsymbol{u} \left\{ \frac{\partial \rho}{\partial t} + \boldsymbol{\nabla} \cdot (\rho \boldsymbol{u}) \right\} + \rho \left(\frac{\partial \boldsymbol{u}}{\partial t} + \boldsymbol{u} \cdot \boldsymbol{\nabla} \boldsymbol{u} \right) \end{aligned} \tag{1.15}$$

The first term on the right side of Eq. (1.15) vanishes by the continuity equation, Eq. (1.3). Next, we introduce the differential operator called the substantial derivative (or Lagrangian derivative) where

$$\frac{D}{Dt} = \frac{\partial}{\partial t} + \boldsymbol{u} \cdot \boldsymbol{\nabla} \tag{1.16}$$

And, considering that $\mu =$ constant, Eq. (1.14) can be written as

$$\frac{D\boldsymbol{u}}{Dt} = -\frac{1}{\rho}\boldsymbol{\nabla} p + 2\nu \boldsymbol{\nabla} \cdot \boldsymbol{S} + \boldsymbol{f} \tag{1.17}$$

where ν is the kinematic viscosity. Here the x-component of the divergence of the strain rate tensor $\boldsymbol{S}$ for incompressible flow is rewritten as

$$\begin{aligned} 2(\boldsymbol{\nabla} \cdot \boldsymbol{S})_x &= \frac{\partial}{\partial x}\left(2\frac{\partial u}{\partial x}\right) + \frac{\partial}{\partial y}\left(\frac{\partial u}{\partial y} + \frac{\partial v}{\partial x}\right) + \frac{\partial}{\partial z}\left(\frac{\partial u}{\partial z} + \frac{\partial w}{\partial x}\right) \\ &= \frac{\partial^2 u}{\partial x^2} + \frac{\partial^2 u}{\partial y^2} + \frac{\partial^2 u}{\partial z^2} + \frac{\partial}{\partial x}\left(\frac{\partial u}{\partial x} + \frac{\partial v}{\partial y} + \frac{\partial w}{\partial z}\right) \\ &= \boldsymbol{\nabla}^2 u + \frac{\partial}{\partial x}(\boldsymbol{\nabla} \cdot \boldsymbol{u}) = \boldsymbol{\nabla}^2 u \end{aligned} \tag{1.18}$$

Applying the same procedures to the y- and z-components,

$$2\nu \boldsymbol{\nabla} \cdot \boldsymbol{S} = \nu \boldsymbol{\nabla}^2 \boldsymbol{u} \tag{1.19}$$

is derived. This allows Eq. (1.17) to be rewritten as

$$\frac{D\boldsymbol{u}}{Dt} = -\frac{1}{\rho}\boldsymbol{\nabla} p + \nu \boldsymbol{\nabla}^2 \boldsymbol{u} + \boldsymbol{f} \tag{1.20}$$

This is the Navier-Stokes equation for incompressible fluid. In the presence of a gravitational field, the body force (force density) $\boldsymbol{f}$ is substituted with $\boldsymbol{g}$ (gravitational acceleration). In Eq. (1.14), the acceleration term is written in the divergence form, while in the following equation

$$\frac{D\boldsymbol{u}}{Dt} = \frac{\partial \boldsymbol{u}}{\partial t} + (\boldsymbol{u} \cdot \boldsymbol{\nabla})\boldsymbol{u} \tag{1.21}$$

the acceleration term is written in the gradient form.

1.1.4. *Indicial notation*

Indicial notation is useful for writing an equation in compact form. For the vector $\boldsymbol{a} = (a_1, a_2, a_3)$ and $\boldsymbol{b} = (b_1, b_2, b_3)$, consider the internal product $\boldsymbol{a}\cdot\boldsymbol{b}=\sum_{i=1}^{3} a_i b_i$ in which i is called an index. To describe this in a compact way, one simple rule can be introduced: "Sum over repeated indices." Applying this rule, the internal product can be written as $\boldsymbol{a} \cdot \boldsymbol{b} = a_i b_i$. Resultingly, we can avoid redundancy by omitting the summation symbol. This form of indicial notation is called the Einstein summation convention.

Writing the coordinate and the velocity component of each coordinate $x = x_1$, $y = x_2$, $z = x_3$, and $u = u_1$, $v = u_2$, $w = u_3$, the mass conservation law (1.3) is written as

$$\frac{\partial \rho}{\partial t} + \frac{\partial(\rho u_j)}{\partial x_j} = 0 \tag{1.22}$$

And the momentum conservation law (1.7) is described as

$$\frac{\partial(\rho u_i)}{\partial t} + \frac{\partial}{\partial x_j}(\rho u_i u_j) = \frac{\partial T_{ij}}{\partial x_j} + \rho f_i \tag{1.23}$$

or

$$\frac{\partial u_i}{\partial t} + u_j \frac{\partial u_i}{\partial x_j} = \frac{1}{\rho}\frac{\partial T_{ij}}{\partial x_j} + f_i \tag{1.24}$$

The stress tensor and the strain rate tensor are

$$T_{ij} = -\delta_{ij} p + 2\mu \left(S_{ij} - \frac{1}{3}\delta_{ij}\frac{\partial u_k}{\partial x_k} \right) \tag{1.25}$$

$$S_{ij} = \frac{1}{2}\left(\frac{\partial u_i}{\partial x_j} + \frac{\partial u_j}{\partial x_i} \right) \tag{1.26}$$

Then the Navier-Stokes equation (1.20) is written as

$$\frac{Du_i}{Dt} = -\frac{1}{\rho}\frac{\partial p}{\partial x_i} + \nu \nabla^2 u_i + f_i \tag{1.27}$$

Equation (1.23) is valid for all indices $i = 1, 2, 3$. This index i is called a free index. On the other hand, index j (j and k in Eq. (1.25)) operates to calculate the total sum and is called a dummy index. In addition, δ_{ij} is the Kronecker delta function:

$$\delta_{ij} = \begin{cases} 1 & i = j \\ 0 & i \neq j \end{cases} \tag{1.28}$$

1.2. Governing Equation for Multiphase Flow

Although the governing equations in the previous section are for single-phase flow, readers may be interested in multiphase flows where gas, liquid and solid-phases coexist. Here, to consider a simple case of multiphase flow, the governing equations of a solid-liquid two-phase flow are shown.

By modeling both the solid and liquid phases as a continuum, the equation of continuity and the equation of motion for the solid-liquid two-phase flow are described as follows:

⟨continuity for liquid phase⟩

$$\frac{\partial}{\partial t}\{\rho_l t(1-c)\} + \nabla \cdot \{(1-c)\boldsymbol{u}_l\} = 0 \tag{1.29}$$

⟨continuity for solid phase⟩

$$\frac{\partial}{\partial t}(\rho_s c) + \nabla \cdot (\rho_s c \boldsymbol{u}_s) = 0 \tag{1.30}$$

⟨momentum for liquid phase⟩

$$\begin{aligned}\rho_l(1-c)\frac{D\boldsymbol{u}_l}{Dt} =& -(1-c)\nabla p_l + \rho_l(1-c)\nu_l\nabla^2\boldsymbol{u}_l \\ &+ \rho_l(1-c)\boldsymbol{g} - \boldsymbol{f}_{ls}\end{aligned} \tag{1.31}$$

⟨momentum for solid phase⟩

$$\rho_s c\frac{D\boldsymbol{u}_s}{Dt} = -c\nabla p_s + \rho_s c\nu_s\nabla^2\boldsymbol{u}_s + \rho_s c\boldsymbol{g} + \boldsymbol{f}_{ls} \tag{1.32}$$

where $\boldsymbol{u}_l$, $\boldsymbol{u}_s$ are the velocity vectors of the liquid and the solid phases; c is the concentration of the solid phase; p_l, p_s are the pressures of the liquid and the solid phases; ρ_l, ρ_s are the densities of the liquid and the solid phases; ν_l, ν_s are the kinematic viscosities of the liquid and the solid phases; $\boldsymbol{g}$ is the gravitational acceleration vector; and $\boldsymbol{f}_{ls}$ is the interaction force between the liquid and the solid phases. The subscripts l and s denote the liquid and solid phases, respectively.

Applying indicial notation, the above-mentioned equation system corresponding to Eqs. (1.29)–(1.32) is written as:

$$\frac{\partial}{\partial l}\{\rho_l(1-c)\} + \frac{\partial}{\partial x_j}\{\rho_l(1-c)u_{lj}\} = 0 \tag{1.33}$$

$$\frac{\partial}{\partial t}(\rho_s c) + \frac{\partial}{\partial x_j}(\rho_s c u_{sj}) = 0 \tag{1.34}$$

$$\rho_l(1-c)\frac{Du_{lj}}{Dt} = -(1-c)\frac{\partial p_t}{\partial x_j} + \frac{\partial}{\partial x_j}\left\{\rho_l(1-c)\nu_l\left(\frac{\partial u_{li}}{\partial x_j} + \frac{\partial u_{lj}}{\partial x_j}\right)\right\} + \rho_l(1-c)g_i - f_{lsi} \quad (1.35)$$

$$\rho_s c\frac{Du_{sj}}{Dt} = -c\frac{\partial p_s}{\partial x_j} + \frac{\partial}{\partial x_j}\left\{\rho_s c\nu_s\left(\frac{\partial u_{si}}{\partial x_j} + \frac{\partial u_{sj}}{\partial x_i}\right)\right\} + \rho_s c g_i + f_{lsi} \quad (1.36)$$

1.3. Governing Equation of Waves

1.3.1. *Potential theory*

Ignoring the viscous term in the Navier–Stokes equation (1.20), the equation of motion for an ideal (inviscid) fluid, namely the Euler equation, is derived.

$$\frac{\partial \boldsymbol{u}}{\partial t} + (\boldsymbol{u}\cdot\boldsymbol{\nabla})\,\boldsymbol{u} = -\frac{1}{\rho}\boldsymbol{\nabla}p + \boldsymbol{f} \quad (1.37)$$

Here, an identical equation of the vector differential operation is applied.

$$\boldsymbol{u}\times(\boldsymbol{\nabla}\times\boldsymbol{u}) = \boldsymbol{\nabla}\left(\frac{|\boldsymbol{u}|^2}{2}\right) - (\boldsymbol{u}\cdot\boldsymbol{\nabla})\boldsymbol{u} \quad (1.38)$$

Introducing vorticity $\boldsymbol{\omega} = \nabla\times\boldsymbol{u}$, then

$$\frac{\partial \boldsymbol{u}}{\partial t} - \boldsymbol{u}\times\boldsymbol{\omega} = -\boldsymbol{\nabla}\left(\frac{|\boldsymbol{u}|^2}{2}\right) - \frac{1}{\rho}\boldsymbol{\nabla}p + \boldsymbol{f} \quad (1.39)$$

is derived. When density is constant ($\rho = \rho_0$), and the external force is a conservative force having a potential χ as follows

$$\boldsymbol{f} = -\boldsymbol{\nabla}\chi \quad (1.40)$$

Eq. (1.39) can be rewritten as

$$\frac{\partial \boldsymbol{u}}{\partial t} - \boldsymbol{u}\times\boldsymbol{\omega} = -\boldsymbol{\nabla}\left(\frac{|\boldsymbol{u}|^2}{2} + \frac{p}{\rho_0} + \chi\right) \quad (1.41)$$

A flow with zero vorticity in the entire domain, namely

$$\boldsymbol{\omega} = \boldsymbol{\nabla}\times\boldsymbol{u} = 0 \quad (1.42)$$

is called an irrotational flow. In an irrotational flow field, the velocity vector is expressed as a gradient of the scalar function Φ as follows:

$$\boldsymbol{u} = \mathrm{grad}(\Phi) = \boldsymbol{\nabla}\Phi \quad (1.43)$$

This scalar function Φ is called the velocity potential. In an irrotational flow (or potential flow), Eq. (1.41) can be rewritten as

$$\boldsymbol{\nabla}\left(\frac{\partial\Phi}{\partial t}+\frac{|\boldsymbol{u}|^2}{2}+\frac{p}{\rho_0}+gz\right)=0 \tag{1.44}$$

In the above equation, $\chi = gz$ (gravitational force) is assumed. Integrating Eq. (1.44), the unsteady Bernoulli equation is derived.

$$\frac{\partial\Phi}{\partial t}+\frac{|\boldsymbol{u}|^2}{2}+\frac{p}{\rho_0}+gz=C(t) \tag{1.45}$$

where $C(t)$ is an integral constant. Furthermore, substituting Eq. (1.43) into the equation of continuity (1.4), the Laplace equation of velocity potential is obtained.

$$\boldsymbol{\nabla}^2\Phi=0 \tag{1.46}$$

The velocity field is determined by solving Eq. (1.46) under the given boundary conditions. Then, the pressure field is evaluated by substituting the velocity-field solution into Eq. (1.45).

In a wave field, the assumption of a potential flow is generally valid, except for the surf zone and around coastal structures where wave energy dissipation is significant. Therefore, Eqs. (1.45) and (1.46) are used as the governing equations for water waves. However, in the surf zone and in areas around coastal structures, the Navier-Stokes equation, or the governing equation of a viscous fluid, must be solved.

1.3.2. *Conventional models for wave transformation analysis*

In deriving a wave equation for waves propagating on the water surface, the x-axis is defined in the direction the waves travel, and the z-axis is defined vertically upward, assuming regular waves propagating in one direction (see Fig. 1.2). The origin of the z-axis is set at the water surface (or the bottom). It is also assumed there is neither motion nor variation in phenomena in the y-direction, a direction that is horizontal and perpendicular to the direction of wave travel. Therefore, the y-derivatives for all variables are zero.

As shown in the previous sub-section, Eq. (1.46) is obtained as the governing equation for a wave flow field by introducing the velocity potential Φ, assuming the flow is incompressible and irrotational. The pressure p

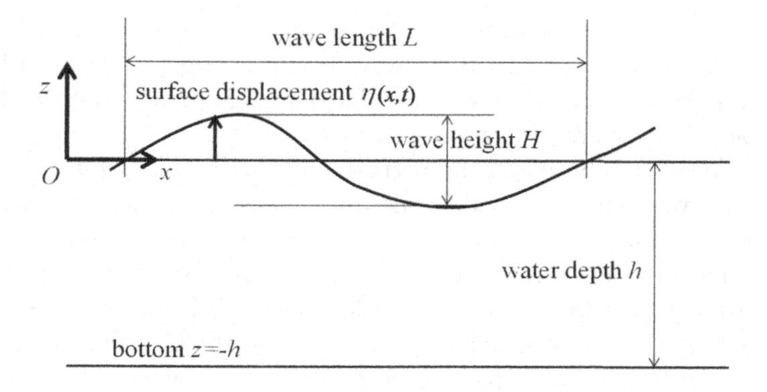

Fig. 1.2. Coordinate system and wave parameters.

should be the same as atmospheric pressure p_0 (= 0 in the gauge pressure) at the water surface. Therefore, Eq. (1.45) can be expressed as

$$\frac{\partial \Phi}{\partial t} + \frac{1}{2}\left(u^2 + w^2\right) + g\eta = C(t) \tag{1.47}$$

at the water surface $z = \eta$. This constraint is called the "dynamic boundary condition at the water surface". If mixing between air and water can be neglected, the water surface is always located on a streamline of the water particle at $z = \eta$. Therefore the following equation for the surface is derived from the equation of streamlines.

$$\frac{\partial \eta}{\partial t} + u\frac{\partial \eta}{\partial x} - w = 0 \tag{1.48}$$

Eq. (1.48) is called the "kinematic boundary condition at the water surface". Other boundary conditions are the kinematic condition at the bottom (the bottom boundary condition) and the conditions for the periodic nature of regular waves (the periodic boundary conditions). Then, Φ of Eq. (1.46) is solved with these conditions. In this sense, solving wave propagation is a typical two-dimensional boundary value problem.

Since Eqs. (1.47) and (1.48) are second order nonlinear differential equations for velocity (= $\nabla\Phi$) and water surface elevation, it is difficult to obtain an analytical solution for Φ under general conditions. A solution is obtained with the small-amplitude wave theory for the linearized equations of (1.47) and (1.48) by omitting the nonlinear terms under the assumption that the change in the surface elevation and the velocity (more specifically, the wave height to wave length ratio and the wave height to water depth ratio) are small (see theoretical solutions for the small amplitude waves, Eqs. (3.74)–(3.77) in §3.2.2). If the wave height is large, Φ obtained by the small-amplitude wave theory will not satisfy Eqs. (1.47) or (1.48) within an allowable error for practical use.

To remedy this problem, the perturbation method is applied to finite-amplitude wave theories. Unknowns are expressed as perturbations expanded around a small parameter ε, and the non-linearity up to a certain order of perturbation terms is considered. Assuming conservative waves, a solution for Φ is uniquely determined for given conditions of water depth h, wave period T (or wave length L) and wave height H. Depending on the value chosen for the small parameter ε, there are some variations such as the Stokes wave theory, described by Eqs. (3.78)–(3.81) in §3.2.2, and the cnoidal wave theory, also by Eqs. (3.82)–(3.85), (see details in Fenton[4]).

1.3.3. *Boussinesq models*

In this sub-section we present an overview of the Boussinesq wave model and its derivative models. Boussinesq models are typical examples of analysis methods for wave propagation and deformation. Also the assumptions inherent in the models and methods to model wave transformation with energy dissipation are briefly explained.

Boussinesq[2] derived a set of equations that describe the weak non-linearity of waves using the perturbation method. Applying the Taylor expansion to the velocity potential Φ at the bottom ($z = -h$) gives

$$\begin{aligned}\Phi = \Phi_b &+ z\left[\frac{\partial \Phi}{\partial z}\right]_{z=-h} + \frac{1}{2}z^2\left[\frac{\partial^2 \Phi}{\partial z^2}\right]_{z=-h} \\ &+ \frac{1}{6}z^3\left[\frac{\partial^3 \Phi}{\partial z^3}\right]_{z=-h} + \frac{1}{24}z^4\left[\frac{\partial^4 \Phi}{\partial z^4}\right]_{z=-h} + \cdots\end{aligned} \tag{1.49}$$

where Φ_b is the value of Φ at the bottom. By further applying the Laplace equation on Eq. (1.49), the following equation is obtained with assumptions that the bottom is horizontal and the flow velocity normal to the bottom is zero.

$$\begin{aligned}\Phi &= \left\{\Phi_b - \frac{1}{2}z^2\frac{\partial^2 \Phi_b}{\partial x^2} + \frac{1}{24}z^4\frac{\partial^4 \Phi_b}{\partial x^4} + \cdots\right\} \\ &\quad + \left\{z\left[\frac{\partial \Phi}{\partial z}\right]_{z=-h} - \frac{1}{6}z^3\frac{\partial^2}{\partial x^2}\left[\frac{\partial \Phi}{\partial z}\right]_{z=-h} + \cdots\right\} \\ &= \left\{\Phi_b - \frac{1}{2}z^2\frac{\partial^2 \Phi_b}{\partial x^2} + \frac{1}{24}z^4\frac{\partial^4 \Phi_b}{\partial x^4} + \cdots\right\}\end{aligned} \tag{1.50}$$

Substituting Eq. (1.50) into Eqs. (1.47) and (1.48) and taking up to the second order for z, the following equations are obtained.

$$\frac{\partial \eta}{\partial t} + \frac{\partial}{\partial x}\left[(h+\eta)\, u_b\right] = \frac{1}{6}h^3 \frac{\partial^3 u_b}{\partial x^3} \tag{1.51}$$

$$\frac{\partial u_b}{\partial t} + u_b \frac{\partial u_b}{\partial x} + g\frac{\partial \eta}{\partial x} = \frac{1}{2}h^2 \frac{\partial^3 u_b}{\partial t\, \partial x^2} \tag{1.52}$$

The term u_b is the horizontal velocity at the bottom. Here, the variation of Φ in the vertical direction is expressed as a quadratic function of z. The right sides of the above equations represent the dispersion of waves, and Eqs. (1.51) and (1.52) agree with the nonlinear long-wave equation if they are taken to be zero.

There are many other variations for these Boussinesq-type equations. Peregrine[11] derived the equations below, which are expressed in terms of velocity $\hat{u}$ averaged over the depth and take into account the bottom slope.

$$\frac{\partial \eta}{\partial t} + \frac{\partial}{\partial x}\left[(h+\eta)\, \hat{u}\right] = 0 \tag{1.53}$$

$$\frac{\partial \hat{u}}{\partial t} + \hat{u}\frac{\partial \hat{u}}{\partial x} + g\frac{\partial \eta}{\partial x} = \frac{1}{2}\frac{\partial}{\partial t}\left[h\frac{\partial^2 (h\hat{u})}{\partial x^2} - \frac{1}{3}h^2\frac{\partial^2 \hat{u}}{\partial x^2}\right] \tag{1.54}$$

In addition, whereas the original Boussinesq equation is basically applied in very shallow water, the modified Boussinesq equations (e.g. Nwogu[10] and Beji and Nadaoka[1]) were proposed with the Φ function chosen to maximize the compatibility of the vertical flow distribution at an arbitrary water depth. It should be remembered that the wave equations are obtained by assuming vertical distribution functions for velocity, and therefore the calculated velocity fields have restricted velocity variations that follow the assumed distribution functions. Boussinesq-type equations are widely used in numerical computations of wave fields (see Fig. 1.3), since they can calculate wave propagation for arbitrary forms of waves, including irregular waves (see Mei[8] and Dingemans[3] for details).

As described above, the wave equations represented by the Boussinesq equation are derived for wave fields without energy dissipation (inviscid and irrotational flows); thus, they are basically not applicable to wave fields with large energy losses associated with turbulence due to wave breaking and interaction with structures. A technique to account for these losses is to add appropriate terms to these equations (not limited to the Boussinesq-type equations) to artificially represent energy dissipation. These additional

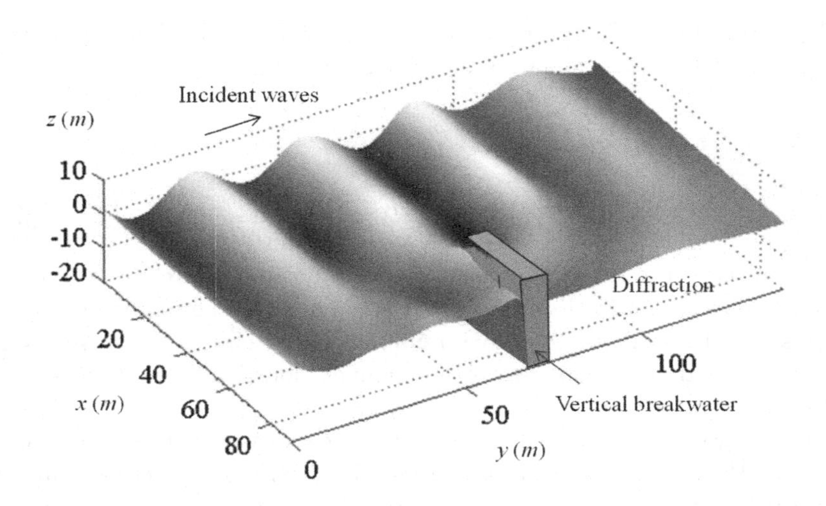

Fig. 1.3. An example of wave transformation around a breakwater head, computed by using the Boussinesq wave model.

terms are classified into three general types: a resistance-type in which the flow resistance is proportional to the flow velocity (e.g. Watanabe and Maruyama[14] for the time-dependent mild slope equation), a dispersive-type where the flow is proportional to the second derivative of velocity in the x-direction (e.g. Kabilling and Sato[5]), and a surface-roller-type that gives a damping force according to the geometry of the water surface (e.g. Schäffer *et al.*[12]).

By adding these additional terms, wave attenuation due to wave breaking can be represented at a practical level. This technique, however, is insufficient for cases where local energy dissipation requires evaluation or more detailed flow information is necessary. For example, assessments and evaluations may not be enough for the wave force and sediment transport by breaking waves, or the fluid force and energy loss around a complex structure or moving body. The wave attenuation calculated with these additional terms is a kind of empirical damping, rather than an attenuation based on physical mechanisms. To begin with, the energy dissipation by wave breaking is beyond the nature of wave motion, thus applying a wave equation that cannot represent the actual mechanism of energy dissipation should have limited use in wave propagation calculations.

Another problem with introducing a wave attenuation term while modeling wave breaking in the surf zone is that the breaking criterion that determines the initial position of breaking needs to be applied to the calculations. There are many breaking criteria in numerical modeling using a wave equation, most of which are based on laboratory experiments and/or field observations of breaking waves. In this context, the breaking criterion proposed by Nadaoka *et al.*,[9] where the wave breaks when "the vertical pressure gradient at the surface of water becomes zero" has a clear physical basis, but prior to adopting it, the wave equation itself should have the capacity to reproduce the water surface profile with enough accuracy. Similarly, applying the shock capturing scheme to wave calculations (e.g. Shimozono *et al.*[13]) also needs a condition such that the instability in the calculation is equivalent to the physical instability of the wave profile.

Some additional problems with using a wave equation to evaluate turbulence include: velocity in the y-direction is not allowed, inviscid and irrotational fluid motions are assumed, and a gradually varying velocity distribution (or a form of the function) is assumed in the vertical direction, in addition to some others. Due to these assumptions, the essence of turbulence, such as being three-dimensional, non-stationary, intermittent, with wide ranges in the frequency and wave-number spectra, are not fully expressed. As a result, turbulent fluid motion, energy dissipation by turbulence, and their influences to the wave field are not properly evaluated. In addition, wave forces and wave overtopping rates are not properly estimated, since irregular changes in water surface elevation due to complex fluid motion and mixing with air (it is even difficult to determine the position of the surface with large amounts of aeration) cannot be directly described.

To overcome these problems, it is necessary to solve the Navier-Stokes equation directly and remove the constraints on how the waves behave. In a numerical wave flume, which is the main object of this text, the Navier-Stokes equation is numerically solved under appropriate boundary conditions (such as the bathymetry and the wave incident boundary). Unlike numerical calculations based on the wave equation, calculations are performed without assuming the wave nature. Thus, a numerical wave flume can consider calculations with gas and solid phases, the interactions between them (that should be properly modeled), as well as turbulence and mixing with air bubbles. Since the calculation load for a numerical wave flume is very high, it is important to choose an appropriate numerical method considering the target objectives and required accuracy.

References

1. Beji S. and Nadaoka K. (1996): A Formal Derivation and Numerical Modelling of the Improved Boussinesq Equations for Varying Depth, *Ocean Engineering*, Vol.23 (8), pp.691-704.
2. Boussinesq J. (1872): Théorie des ondes et des remous qui se propagent le long d'un canal rectangulaire horizontal, en communiquant au liquide contenu dans ce canal des vitesses sensiblement pareilles de la surface au fond, *Journal de Mathématique Pures et Appliquées*, Deuxième Série 17: pp.55-108.
3. Dingemans M. W. (1997): Wave Propagation over Uneven Bottoms, *Advanced Series on Ocean Engineering* 13, Part 2, Chapter 5, World Scientific.
4. Fenton J. D. (1990): Nonlinear wave theories, *The Sea - Ocean Engineering Science*, Part A, B. Le Méhauté & D. M. Hanes (eds), Vol. 9, Wiley, New York, pp.3-25.
5. Kabiling M. B. and Sato S. (1993): Two-dimensional Nonlinear Dispersive Wave-Current Model and Three-dimensional Beach Deformation Model, *Coastal Engineering in Japan*, JSCE, Vol. 36, No.2, pp.195-212.
6. Lamb H. (1932): *Hydrodynamics*, Cambridge University Press, p.738.
7. Landau L. D. and Lifshitz E. M. (1987): *Fluid Mechanics*, Pergamon Press, p.551.
8. Mei C. C. (1989): The Applied Dynamics of Ocean Surface Waves, *Advanced Series on Ocean Engineering* 1, Chapter 11, World Scientific.
9. Nadaoka K., Ono O. and Kurihara H. (1997): Near-Crest Pressure Gradient of Irregular Water Waves Approaching to Break, *Proc. of Coastal Dynamics* '97, ASCE, pp.255-264.
10. Nwogu O. (1993): Alternative Form of Boussinesq Equations for Nearshore Wave Propagation, *Journal of Waterway, Port, Coastal and Ocean Engineering*, ASCE, Vol.119 (6), pp.618-638.
11. Peregrine D. H. (1967): Long waves on a beach, *Journal of Fluid Mechanics*, Vol.27 (4), pp.815-827.
12. Schäffer H. A., Deigaard R. and Madsen P. A. (1992): A Two-Dimensional Surfzone Model Based on the Boussinesq Equations, *Proc. 23rd Int. Conf. on Coastal Eng.*, ASCE, pp.576-589.
13. Shimozono T., Kusano M., Tajima Y. and Sato S. (2007): Application of Shock Capturing Schemes for Deformation of Neashore Periodic Waves, *Annual Journal of Coastal Engineering*, JSCE, Vol.54, pp.26-30 (in Japanese).
14. Watanabe A. and Maruyama K. (1986): Numerical Modeling of Nearshore Wave Field under Combined Refraction, Diffraction and Breaking, *Coastal Engineering in Japan*, JSCE, Vol.29, pp.19-39.

Chapter 2

Turbulence Model

Nobuhito Mori, Yasunori Watanabe, Takayuki Suzuki and Akio Okayasu

In flow fields, turbulence plays various roles. It produces shear stresses exerted as a Reynolds stress that change the profiles of mean flow, increases the diffusion of substances mixed in a fluid, and dissipates the predominant flow energy through an energy cascade-down process. Turbulence often dominates and characterizes nearshore wave fields involving nearshore currents, breaking waves, and flow fields near bottom surfaces and around structures.

In this chapter, the major features of turbulent flows in wave fields and the computational methodologies to model them are first introduced (§2.1). Standard approaches of the Reynolds Averaged Navier–Stokes (RANS) and Large Eddy Simulation (LES) models are interpreted in §2.2 and 2.3, respectively. Then, in §2.4, these models are applied to several examples of wave-breaking turbulence in the surf zone. Advantageous features and limitations of the current turbulence models are also discussed to provide insight into the directions of future wave-breaking research.

2.1. Nearshore Wave Fields and Turbulence

2.1.1. *Definition of turbulence in wave fields*

If there is no loss of energy or exchange of momentum due to turbulence, the flow under pure wave motion can be treated as potential flow except near boundaries such as the bottom, a wall, and the water surface. In this case, with potential flow as described in §1.3, various wave equations can be derived as approximate solutions of the Laplace equation, the governing equation for incompressible potential flows. Then the entire wave field can be solved analytically or numerically with one of the wave equations and specified boundary conditions, such as a boundary condition for incident waves and bathymetry. The irrotational assumption used in wave models never expresses rotational turbulent features and resulting dissipation. To describe wave decay due to wave breaking or interaction with structures, an additional energy dissipation term needs to be included in the equations to reduce an appropriate amount of energy in the wave field. This energy dissipation term generally has empirical factors that are intentionally adjusted to achieve proper wave height changes observed in laboratory tests or field surveys. Therefore these models cannot account for various phenomena that are not attributed to wave motion, such as large-scale vortex motions, momentum exchange by turbulence, and temporal and spatial variabilities of diffusion.

The numerical wave flume (NWF) is a water wave simulator with a flow model that can describe the generation and dissipation of turbulence based on its mechanism[a], and thus it is an advantageous tool to properly evaluate the temporal and spatial variations of turbulence and the resultant flow changes described above. Before we discuss turbulence calculated in the NWF, the definition of turbulence under wave fields should be made clear.

In steady conditions, turbulence can generally be defined as the deviation from the time-averaged flow velocity. For unsteady wave motion, however, there are several definitions for turbulence. Some typical definitions are: (i) deviation from the equi-phase-mean velocity, which is obtained by averaging the velocity at the same phase over a number of waves; (ii) signature remains after extraction through a time filter or a space filter; (iii) correlation with physical quantities, such as surface elevation, to represent fluctuations by minor wave motions. For definition (i), it is easy to understand an analogy of turbulence in steady flow, but it can only be applied to regular (monochromatic) wave trains. Organized fluid motion due to large-scale vortices that repeatedly occur at the same phase of every wave, as observed in breaking waves and near coastal structures, may be included in the mean flow and therefore excluded from this definition of turbulence. Since these types of vortices do not indicate any definitive wave nature, they should be classified as turbulence. For definition (ii), there is a problem of arbitrariness with the filter design (such as the cut-off frequency or wave number) as well as a problem with whether the turbulence can be properly separated by a static (time-and space-independent) filter. In definition (iii),

[a]A numerical wave tank (NWT) has been commonly used to define a wave simulator with a boundary integral method or depth-integrated wave equation model to compute surface elevation. To distinguish both the methodology and governing equations from those used in a NWT, we will use NWF rather than NWT in this book.

turbulence is defined as a quantity that has no correlation with surface elevation change, which represents irregularities involved with wave motion. Also in this method, the correlation filter (or transfer function) is selected arbitrarily, and the surface fluctuation itself is affected by large-scale turbulence, which may produce variable features that deviate from turbulence statistics. Each of these definitions for turbulence has both advantages and disadvantages, and the best definition depends on the problem.

The Navier–Stokes equation can be solved directly with suitable multidimensional numerical methods, an approach that is called a direct numerical simulation (DNS). If DNS is applied to complex turbulent flow, the grid size should be small enough so that the momentum exchange due to turbulence is negligible when compared to the exchange by molecular viscosity. Since DNS is very costly, numerical models with larger grid sizes and time intervals are adopted and widely used with appropriate turbulence models. In these models, momentum exchange in the grid-size resolution is evaluated with calculated turbulence quantities, and generation, advection and dissipation of turbulence are modeled according to the physical processes. Turbulence definition (ii) is applied in the models that provide statistical quantities such as turbulence intensity, dissipation, and diffusivity. While no turbulent velocities are computed for resolutions finer than the grid size, turbulent motions with variations larger than the grid scale are explicitly computed. Therefore a scale separation of turbulent flow, constrained by the length of the grid size (in the space domain) or the time interval (in the time domain), is an important issue when discussed in this context.

2.1.2. *Overviews of turbulence models in numerical wave flumes*

As already mentioned, in the conventional wave field calculation using a wave equation, turbulence was not explicitly determined. In the NWF, interactions between the computed turbulence and main flow field using an appropriate turbulence model are reasonably assessed. As also indicated, the definition of turbulence or the method of turbulence separation has not been uniquely identified, especially in unsteady wave fields. It is common, however, to define turbulence in calculations as the deviation from the main flow that is defined with representative length or time scales.

Two typical turbulence models based on different definitions of turbulence are widely used in NWFs. In this chapter these models are explained: the Reynolds averaged model and the Large Eddy Simulation model. The appropriate model for a particular case should be determined in terms of the target flow conditions and restrictions on calculations. For all cases, common features of turbulence include: non-stationary aspect in time and space (stochastic variability), nonlinearity, and continuity in the spectra for the wave number and frequency domains. A turbulent flow model that ensures the above features should be incorporated into the flow computation for nearshore wave fields where interactions between the main flow and turbulence are enhanced. These effects of turbulence are never estimated by any conventional wave equation model, and the computational costs are expensive when compared to the wave equation model.

Additionally, errors due to numerical diffusion and numerical viscosity result from discretization and approximation (see details in §3.1); these errors are inevitable in numerical simulations. To evaluate the reproducibility of the simulation, it should be noted that the numerical diffusion and viscosity are different from turbulent diffusion and eddy viscosity calculated by a turbulence model. In other words, agreement between a numerical calculation and a laboratory experiment or field observed result does not always indicate a capable turbulence model; the result also depends highly on the discretized error and numerical accuracy of the numerical scheme used. Extra caution should be taken to verify the computed result. Comparing a single physical quantity is inadequate to ensure a reliable computation, and therefore it is necessary to examine and analyze the entire flow field, referring to multiple physical quantities[b].

2.2. Reynolds Averaged Model

2.2.1. *Basic concept*

A turbulent flow involves complex fluid motion fluctuating at wide-ranging temporal and spatial scales. Assuming that any fluid motion can be separated into a main flow and a disturbance flow, a Reynolds decomposition is used to separately assess the coherent characteristics of the main flow and the white-noise-like disturbance flow. That is, the instantaneous flow velocity $\boldsymbol{u}$ can be described as the summation of the mean component $\boldsymbol{U}$ and the fluctuation $\boldsymbol{u}'$. Based on the Reynolds decomposition, there are

[b]In general, surface elevation exhibits insensitive variations to the local fluid velocity beneath the surface. Therefore, correlating the surface elevation with an experimental one does not directly ensure the correct computation of fluid flow and turbulence. The computed velocity and turbulence statistics should be compared to validate the NWF model used.

three different definitions of statistical mean: (i) temporal mean, (ii) spatial mean, and (iii) ensemble mean. The definition that is most suitable to the statistical property of the flow should be chosen, while in standard turbulence theory, the ensemble mean definition is traditionally used.

Flow in the surf zone contains multiple unsteady modes: an ocean wave component, transitional large-scale vortices due to wave splashing, and unsteady small-scale turbulence. Therefore, for this type of flow, the statistical measure is not straightforwardly determined and is interesting to consider[c].

The temporal mean may be easier to understand after a general framework of the separation of flows is presented. Therefore, it is presented in the latter part of this section. Following standard methods in fluid mechanics, the Reynolds decomposition is performed on the basis of a short-term moving average for flow velocity.

$$\boldsymbol{u}(\boldsymbol{x},t) = \boldsymbol{U}(\boldsymbol{x},t) + \boldsymbol{u}'(\boldsymbol{x},t) \tag{2.1}$$

$$\boldsymbol{U}(\boldsymbol{x},t) = \frac{1}{T}\int_t^{t+T} \boldsymbol{u}(\boldsymbol{x},t)dt \tag{2.2}$$

$$\overline{\boldsymbol{u}'}(\boldsymbol{x},t) = \frac{1}{T}\int_t^{t+T} \boldsymbol{u}'(\boldsymbol{x},t)\,dt = 0 \tag{2.3}$$

where T is the representative time scale of the moving average and $\boldsymbol{x}$ and t are spatial and temporal variables respectively. This operation can be applied to all velocity components and pressure p (see Fig. 2.1).

The correlation between velocity components u_i and u_j $(i, j = 1, 2, 3)$ follows

$$\overline{u_i u_j} = \overline{U_i U_j} + \overline{U_i u'_j} + \overline{u'_i U_j} + \overline{u'_i u'_j} = \overline{U_i U_j} + \overline{u'_i u'_j} \tag{2.4}$$

Here, $\overline{U_i u'_j} = \overline{u'_i U_j} = 0$ because the averaged random fluctuation is zero, $\overline{u'_i} = \overline{u'_j} = 0$. But generally, $\overline{u'_i u'_j} \neq 0$ and this term indicates a correlation between the turbulence components u'_i and u'_j.

The Reynolds-Averaged Navier–Stokes equation can be derived by applying the Reynolds decomposition to the continuity equation Eq. (1.4) and the Navier Stokes equation Eq. (1.20) and taking the average of both of the

[c]Since water wave motion is well approximated by unsteady irrotational flow, the scale separation of the flow needs to be carefully considered. Accordingly, if the time average of the wave field is taken over a time frame that is longer than the wave period, the irrotational wave components are also estimated to be fluctuating velocities as well as real turbulence components. The irrotational flows are an obvious deviation from the definition of turbulence.

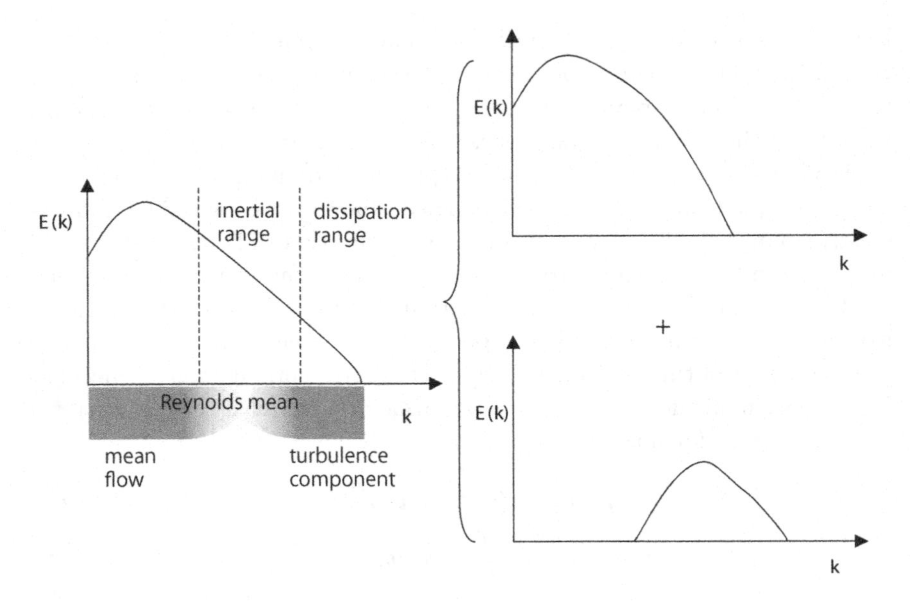

Fig. 2.1. Illustration of the wave number spectrum of flow velocity and related scales ik: wave number).

equations:

$$\frac{\partial U_i}{\partial x_i} = 0 \tag{2.5}$$

$$\frac{\partial U_i}{\partial t} + U_j \frac{\partial U_i}{\partial x_j} = -\frac{1}{\rho}\frac{\partial P}{\partial x_i} + \nu \frac{\partial S_{ij}}{\partial x_j} - \frac{\partial \overline{u_i' u_j'}}{\partial x_j} + F_i \tag{2.6}$$

$$S_{ij} = \left(\frac{\partial U_i}{\partial x_j} + \frac{\partial U_j}{\partial x_i}\right) \tag{2.7}$$

where $\boldsymbol{u} = u_i = U_i + u_i',\ p = P + p'$, S_{ij} is the strain tensor, and F_i is the body force, such as gravitational acceleration. The third term on the right hand side of Eq. (2.6) involves the correlation of turbulent components in Eq. (2.4), which indicates the interactions between mean flow and disturbance flow, or, momentum transfer due to turbulence. Since this term is not determined by any variables in the equation system of Eqs. (2.5)–(2.7), an additional assumption or equation is needed to close the system. This term is denoted the Reynolds stress, τ_{ij}

$$\tau_{ij} = -\rho\overline{u_i' u_j'} \tag{2.8}$$

The Reynolds stress needs to be expressed with main flow parameters via some simple models to close the system. Since the Reynolds stress incorporates the second-order correlation of turbulence components, further higher-order correlation is required to explicitly determine τ_{ij}; that is, the equation system will never close without some appropriate approximations, which is known as a turbulence closure problem. The set of equations that comprise the Reynolds-Averaged Navier–Stokes equation and turbulence model is simply called a RANS model.

2.2.2. *Reynolds stress and turbulence energy*

There are major two types of RANS models: an eddy viscosity model and a stress equation model. The former model assumes mechanical contributions of τ_{ij} described by an expression similar to laminar shear, while the latter type includes modeling the transport equation for the Reynolds stress.

There have been many engineering applications of the eddy viscosity model during the past three decades. The eddy viscosity representation of the Reynolds stress is

$$-\overline{u_i'u_j'} = \nu_T S_{ij} - \frac{2}{3}\delta_{ij}k \tag{2.9}$$

where $k = \frac{1}{2}\overline{u_i'u_i'}$ is the turbulent kinetic energy, and $\nu_T = \mu_T/\rho$ is the eddy viscosity. This simple model, based on the analogy of momentum transfers due to turbulence and molecular viscosity, is called a Boussinesq approximation. In this model, the influence of turbulence on the mean flow is assumed to be proportional to the velocity gradient perpendicular to the stream line, similar to molecular viscosity.

The transport equation for turbulent energy k can be expressed as

$$\frac{\partial k}{\partial t}+U_j\frac{\partial k}{\partial x_j} = -\overline{u_i'u_j'}\frac{\partial U_j}{\partial x_j}-\nu\overline{\frac{\partial u_i'}{\partial x_k}\frac{\partial u_i'}{\partial x_k}}+\frac{\partial}{\partial x_j}\nu\frac{\partial k}{\partial x_j}-\frac{\partial}{\partial x_j}\left(\frac{1}{2}\overline{u_i'u_i'u_j'}+\frac{1}{\rho}\overline{p'u_j'}\right) \tag{2.10}$$

The second term on the right hand side indicates the energy dissipation due to turbulence:

$$\varepsilon = \overline{\frac{\partial u_i'}{\partial x_k}\frac{\partial u_i'}{\partial x_k}} \tag{2.11}$$

The turbulent diffusion term, or the fourth term on the right hand side of Eq. (2.10) (the triple correlation of turbulent velocities plus the correlation between pressure fluctuation and turbulence), is assumed to be modeled by the gradient of the turbulent energy:

$$\frac{1}{2}\overline{u_i'u_i'u_j'} + \frac{1}{\rho}\overline{p'u_j'} = -\frac{\nu_T}{\sigma_k}\frac{\partial k}{\partial x_j} \tag{2.12}$$

where σ_k is the turbulence Prandtl number. This assumption has been validated for channel flow, as computed by DNS.[16] Substituting Eq. (2.11) and Eq. (2.12) into Eq. (2.10), we have

$$\frac{\partial k}{\partial t} + U_j \frac{\partial k}{\partial x_j} = \frac{1}{\rho}\tau_{ij}\frac{\partial U_j}{\partial x_j} - \varepsilon + \frac{\partial}{\partial x_j}\left[\left(\nu + \frac{\nu_T}{\sigma_k}\right)\frac{\partial k}{\partial x_j}\right] \tag{2.13}$$

where τ_{ij} in the first term on the right hand side can be determined with Eq. (2.9) and the computed mean flow gradients and turbulent energy. However, the second term on the right hand side, the energy dissipation term ε, is not explicitly determined, and therefore an additional model is required to close the equation system.

Many turbulence models for Eq. (2.9) and Eq. (2.13) have been developed, each with different model assumptions that are empirically designed to be consistent with specific types of flow. Therefore, the computed results for arbitrary untrivial flows may depend on the model assumptions. The chosen turbulence model should be carefully validated for the specific application, comparing the experimental or DNS results.

General descriptions of turbulence models are summarized in the following subsection.

2.2.3. *Turbulence in the RANS model*

In the eddy viscosity model of Eq. (2.9), the turbulent energy k is an essential parameter to determine ν_T; however, there are arbitrary choices for the additional parameter(s) for ν_T. There are three major ideas behind choosing the scaling parameter (or explanatory variable) for turbulence modeling, and these ideas stem from dimensional analysis.

Assuming that the eddy viscosity can be expressed as a combination of the turbulence energy k and the representative turbulence scale l,[20]

$$\nu_T \propto k^{1/2}/l \qquad \varepsilon \propto k^{3/2}/l \tag{2.14}$$

The assumption of eddy viscosity as a function of k and l is applied to a single equation model, which will be explained later.

Assuming the turbulent energy k and turbulent dissipation ε as scaling parameters, the eddy viscosity can be expressed as[3]

$$\nu_T \propto k^2/\varepsilon \qquad l \propto k^{3/2}/\varepsilon \tag{2.15}$$

Since the energy dissipation is closely related to the evolution of the turbulent energy, which is trivial from Eq. (2.13), this combination of parameters

k and ε is physically rational, and therefore widely applied as a two-equation model.

Alternatively, assuming that the eddy viscosity is expressed by the turbulence energy k and the specific dissipation rate ω,[11,21,29]

$$\nu_T \propto k/\omega \qquad l \propto k^{1/2}/\omega \qquad \varepsilon \propto \omega k \tag{2.16}$$

ω has a dimension of $[T^{-1}]$, the physical meaning may be interpreted as either energy dissipation per unit time and unit volume, or characteristic time scale of turbulence. The eddy viscosity represented by k and ω is also widely used, especially in the area of geophysical fluid modeling such as atmospheric and ocean currents.

2.2.3.1. *One equation model*

The eddy viscosity from Eq. (2.14) may be appropriate when the mean flow velocity profile is roughly estimated, *i.e.* the length scale l can be explicitly given. For this case, if the energy dissipation is expressed as $\varepsilon = C_D k^{3/2}/l$ with the constant C_D, Eq. (2.13) becomes

$$\frac{\partial k}{\partial t} + U_j \frac{\partial k}{\partial x_j} = \frac{1}{\rho}\tau_{ij}\frac{\partial U_j}{\partial x_j} - C_D \frac{k^{\frac{3}{2}}}{l} + \frac{\partial}{\partial x_j}\left[\left(\nu + \frac{\nu_T}{\sigma_k}\right)\frac{\partial k}{\partial x_j}\right] \tag{2.17}$$

$$\nu_T = k^{\frac{1}{2}} l \tag{2.18}$$

This formulation is well known as the one equation model for the RANS model. The eddy viscosity Eq. (2.18) is applied to the Reynolds equation Eq. (2.6) with Eq. (2.9). Although this model is very simple and easy to use, the turbulence scale l should be known in advance. Thus, the application is limited to specific, trivial flows.

2.2.3.2. *Two equation model*

A two equation model using an additional equation to estimate the eddy viscosity ν_T has been proposed to compute more general flows. In this model, as seen in Eq. (2.15) and Eq. (2.16), the general behavior of turbulent fluid is parameterized by the energy dissipation rate ε, or the specific dissipation rate ω, as well as k. The most popular two equation model in engineering applications is known as the $k - \varepsilon$ model.[10,13] The transport equation for ε is written as[7]

$$\frac{\partial \varepsilon}{\partial t} + U_j \frac{\partial \varepsilon}{\partial x_j} = C_{\varepsilon 1}\frac{\varepsilon}{k}\tau_{ij}\frac{\partial U_j}{\partial x_j} - C_{\varepsilon 2}\frac{\varepsilon^2}{k} + \frac{\partial}{\partial x_j}\left[\left(\nu + \frac{\nu_T}{\sigma_\varepsilon}\right)\frac{\partial \varepsilon}{\partial x_j}\right] \tag{2.19}$$

$$\nu_T = C_\mu k^2/\varepsilon \tag{2.20}$$

where the constants $C_{\varepsilon 1} = 1.44$, $C_{\varepsilon 1} = 1.92$, $\sigma_\varepsilon = 1.3$, and $C_\mu = 0.09$, which are empirically adjusted to be consistent with free shear flows. In Eq. (2.13), $\sigma_k = 1.0$ is normally used. The system of equations Eq. (2.13), Eq. (2.19) and Eq. (2.20) is applicable to fully developed turbulence at high Reynolds numbers and is called a standard (high Reynolds number type) $k - \varepsilon$ model. The standard $k - \varepsilon$ model is known to give erroneous results for anisotropic flows near a rigid boundary. And, a low Reynolds number type $k - \varepsilon$ model has been also developed, where $C_{\varepsilon 2}$ in Eq. (2.19) and C_μ in Eq. (2.20) are determined as a function of the Reynolds number.[10]

As mentioned previously, the constants in Eq. (2.13), Eq. (2.19) and Eq. (2.20) are empirically determined using specific experimental results; there is no guarantee that the model can reproduce arbitrary, turbulent dynamics.[13] The $k-\varepsilon$ model has also been used to compute wave turbulence in the surf zone, which will be shown at the end of this chapter.

2.2.3.3. *Generic Length Scale (GLS) model*

As mentioned above, many other two equation models have been proposed in addition to the $k - \varepsilon$ model. To summarize the various types of the two equation models, Umlauf *et al.*[27] generalized the model by introducing a general function ψ with turbulence energy k. This model is called the Generic Length Scale (GLS) model, and also uses the transport equation for k, Eq. (2.13). The energy dissipation ε is modeled as

$$\varepsilon = (c_\mu^0)^{3+\frac{q}{n}} k^{\frac{3}{2}+\frac{m}{n}} \psi^{-\frac{1}{n}} \tag{2.21}$$

where q, m, and n are constants. With values of q, m and n, Eq. (2.21) becomes different for the various types of two equation models such as $k-\varepsilon$, $k-\omega$, and so on. The constant c_μ^0 is the stability function and is empirically given on the basis of a logarithmic assumption.

The GLS model uses the general function ψ following the transport equation:

$$\frac{\partial \psi}{\partial t} + U_i \frac{\partial \psi}{\partial x_i} = \frac{\psi}{k}\left(C'_{\varepsilon 1} \frac{\varepsilon}{k} \tau_{ij} \frac{\partial U_j}{\partial x_j} - C'_{\varepsilon 2} \varepsilon \right) + \frac{\partial}{\partial x_j}\left[\left(\nu + \frac{\nu_T}{\sigma_\varepsilon} \right) \frac{\partial \varepsilon}{\partial x_j} \right] \tag{2.22}$$

where the constants $C'_{\varepsilon 1}$ and $C'_{\varepsilon 2}$ are determined to adjust the Karman constant and the experimental results of isotropic turbulence.

The general function is defined by

$$\psi = (c_\mu^0)^q \, k^m \, l^n \tag{2.23}$$

$$l = (c_\mu^0)^3 k^{\frac{3}{2}} \, \varepsilon^{-1} \tag{2.24}$$

Umlauf *et al.*[27] formulated the relationship between ψ, ε, kl and ω systematically. The GLS model can be applied to any two equation models with the values of q, m, and n. For example, Eqs. (2.22) and (2.23) with $(q, m, n) = (0, 1, 1)$ become a Mellor–Yamada level 2.5 scheme[17] which is often used for geophysical flows.

Fundamental features of the RANS model with Reynolds decomposition and scale separation of the Navier–Stokes equation are interpreted in this section. The RANS model is advantageous in terms of computational cost because relatively large computational grids can be taken to be identical in scale with the mean flow. Therefore, the RANS model has been used in engineering applications requiring large computational domains. Separation of scale under Reynolds decomposition is evaluated assuming perfect separation of the mean flow and the turbulence scale; that is, both scales are assumed to never overlap. However, it is often difficult to adhere to this fundamental premise for complex flow fields with a wide range of turbulent flow. The explicit scale separation of turbulent flow is used in a Large Eddy Simulation (LES), which is explained in the next section.

2.3. Large Eddy Simulation

2.3.1. *Basic concept*

The RANS approach provides simple models to describe statistical contributions of the velocity fluctuations over a range of wave-numbers to the statistically averaged velocity. With LES, the dynamics of fluid motion is dealt with separately at large- and small-scales with a boundary cut-off length Δ. While the large-scale motion is explicitly computed, contributions from small-scale fluctuations associated with wave-numbers higher than $k_c = \pi/\Delta$, a so-called cut-off wave-number, are statistically modeled. This scale separation is achieved with a spatial filtering of the instantaneous velocity. Accordingly, in physical space, the velocity is separated into a filtered velocity without fluctuations less than Δ, and a residual smaller-scale velocity (see Fig. 2.2). The former is also called a resolved or grid-scale (GS) velocity as Δ is normally used to be equivalent with the computing grid width; and the latter subgrid (SG) velocity represents fluctuations unresolved on the discrete grid system. This operation works in wave-number space as a low-pass filter to eliminate an energy spectrum in a range higher than k_c (see Fig. 2.3). The passed spectrum is resolved for explicit computing while the residual spectrum higher than k_c is described by a statistical subgrid model.

LES can explicitly compute large-scale unsteady motion and is expected to be more accurate and reliable than RANS simulations that often give diffusive outputs for such unsteady vortical flows. It is also advantageous to compute coherent structures that are locally governed by resolved fluid dynamics, for which RANS simulations can produce significant errors. On the other hand, LES generally require more computational cost than RANS simulations; LES require sufficiently fine computing grids to define the maximum length scale of subgrid fluctuations that will be statistically described by the energy spectrum in an inertial subrange, that is, Δ should be taken to be comparable with Kolmogolov's length scale.

Despite remarkable advancements in turbulence modeling of mixing layers and boundary layers (e.g. Piomelli[19]), turbulence at the water surface is still difficult to model with either LES or RANS simulations because anisotropic turbulence is at a discontinuous state for density, viscosity and surface tension; this modifies the turbulence statistics across an interface. Any physical responses and turbulent interactions at the interface may not be understood, and thus no available model has yet been considered. Especially with LES, the filtering operation in the vicinity of the surface should be carefully examined (Labourasse *et al.*[12]). These challenging issues may substantially relate to nearshore wave dynamics, especially those processes that occur during wave breaking; therefore, in this section, some common subgrid models are explained as a first step to understand the theoretical backgrounds of LES and their computational limitations. Readers should refer to Sagaut[22] and Meneveau and Katz[18] for discussions on a variety of models.

2.3.2. *Filtering operation*

The filtering operation for a variable $\psi(\boldsymbol{x}, t)$ is defined as a convolution product with a filter function $\boldsymbol{G}(\boldsymbol{x} - \boldsymbol{x}')$ that is spatially integrated over the domain:

$$\overline{\psi(\boldsymbol{x}, t)} = \int \boldsymbol{G}(\boldsymbol{x} - \boldsymbol{x}')\psi(\boldsymbol{x}', t)d^3\boldsymbol{x}' \tag{2.25}$$

where the filter function fulfills the normalized condition

$$\int \boldsymbol{G}(\boldsymbol{x}')d^3\boldsymbol{x}' = 1 \tag{2.26}$$

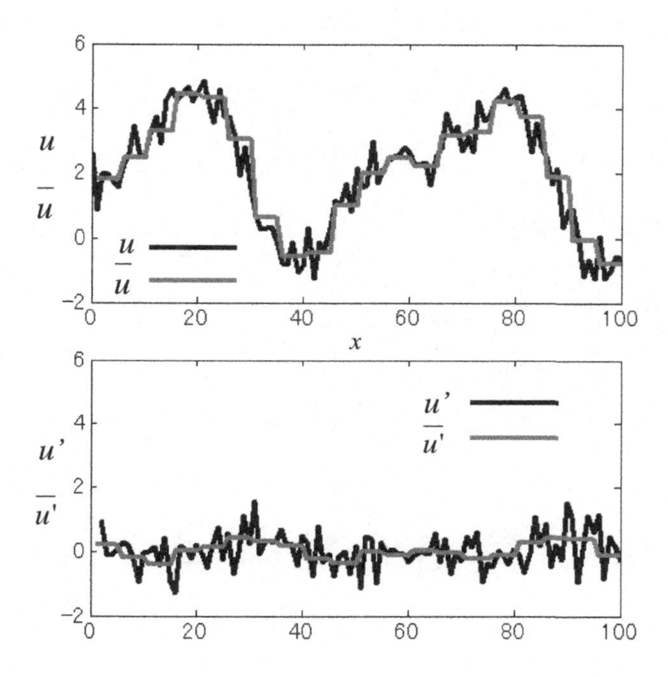

Fig. 2.2. Schematic representation of instantaneous velocity u, resolved (GS) velocity $\overline{u}$, residual (SG) velocity u' and filtered residual velocity $\overline{u'}$.

The residual between the instantaneous quantity ψ and the resolved quantity $\overline{\psi}$ is defined as

$$\psi'(\boldsymbol{x},t) = \psi(\boldsymbol{x},t) - \overline{\psi(\boldsymbol{x},t)} \tag{2.27}$$

The filtered residual is not zero, unlike the mean fluctuations defined in the Reynolds decomposition explained in §2.1 (see Fig. 2.2):

$$\overline{\psi'(\boldsymbol{x},t)} \neq 0 \tag{2.28}$$

The homogeneous filter function used in LES is designed to satisfy the following conditions to manipulate the differential equations to be solved.

(1) The filtered constant is unchanged, $\overline{a} = a$, which is trivial from Eq. (2.26)

(2) Linearity of the filtered variables is preserved, $\overline{\psi_1 + \psi_2} = \overline{\psi_1} + \overline{\psi_2}$

(3) The filtering operation commutes with differentiation, $\overline{\dfrac{\partial \psi}{\partial x}} = \dfrac{\partial \overline{\psi}}{\partial x}$

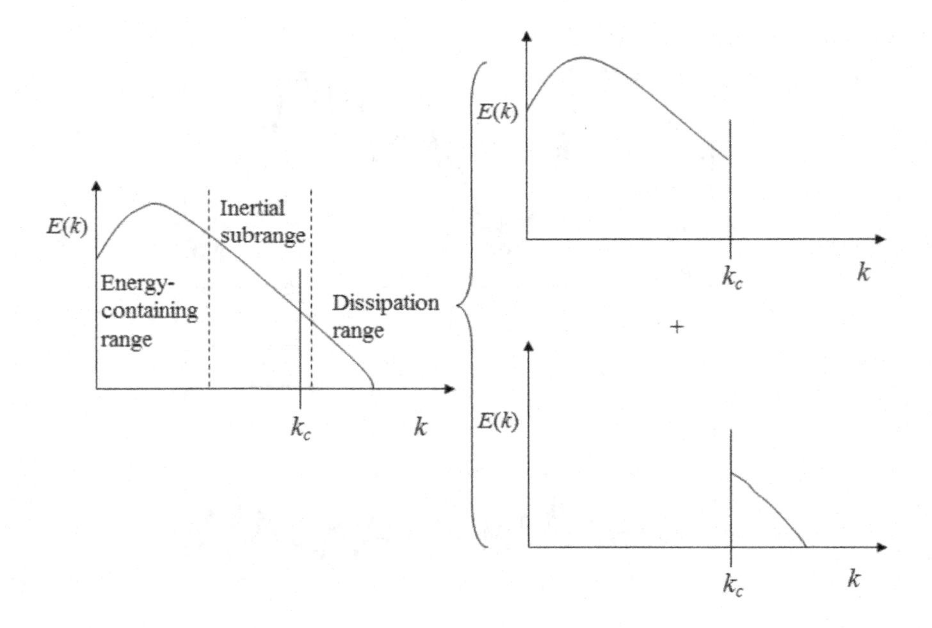

Fig. 2.3. Schematic representation of scale separation; energy spectrum for resolved velocity when $k < k_C$(top), and subgrid energy spectrum to be modeled when $k > k_C$(bottom).

While a filter function with compact support in both the physical and wave-number spaces should be chosen to attain complete scale separation, there is no such ideal function. Therefore, many selective functions have been proposed. Some common filters used in LES are presented below.

A top-hat filter that provides a spatial average over $\boldsymbol{x}' - \Delta/2 < \boldsymbol{x} < \boldsymbol{x}' + \Delta/2$ is written as

$$\boldsymbol{G}(\boldsymbol{x}) = \frac{1}{\Delta}\boldsymbol{H}(\Delta/2 - |\boldsymbol{x}|) \tag{2.29}$$

where

$$\boldsymbol{H}(\boldsymbol{\xi}) = \begin{cases} 1 \ (\boldsymbol{\xi} \geq 0) \\ 0 \ (\boldsymbol{\xi} < 0) \end{cases} \tag{2.30}$$

A Fourier representation of the filtered variable is also defined by multiplying a Fourier transform of $\psi(\boldsymbol{x})$, $\hat{\psi}(\boldsymbol{k})$, and a transfer function $\tilde{\boldsymbol{G}}(\boldsymbol{k})$ (a spectrum of $\boldsymbol{G}(\boldsymbol{x})$):

$$\hat{\overline{\psi}}(\boldsymbol{k}) = \tilde{\boldsymbol{G}}(\boldsymbol{k})\hat{\psi}(\boldsymbol{k}) \tag{2.31}$$

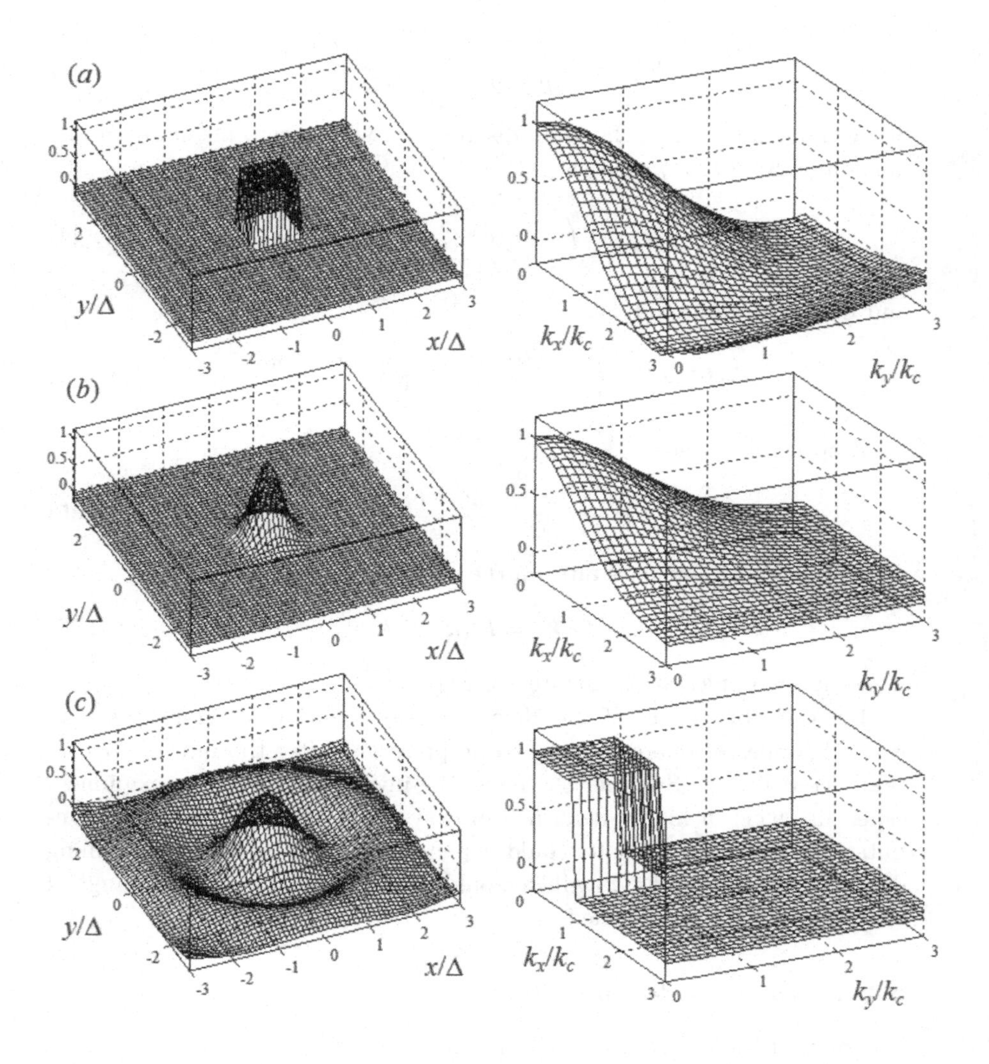

Fig. 2.4. Top-hat filter (a), Gaussian filter (b) and spectrum filter (c); filter kernel in a physical space (left panel), transfer function in a Fourier space (right panel).

The transfer function takes the value of one at $\boldsymbol{k} = 0$ and normally decays with $\boldsymbol{k}$, which operates the amplitude of $\hat{\psi}(\boldsymbol{k})$ in a range of the filter support; a large scale variable is preserved when $\tilde{\boldsymbol{G}}(\boldsymbol{k}) = 1$ and a small scale variable is filtered out when $\tilde{\boldsymbol{G}}(\boldsymbol{k}) = 0$. Thus, the transfer function for the top-hat filter is

$$\tilde{\boldsymbol{G}}(\boldsymbol{k}) = \frac{\sin(\boldsymbol{k}\Delta/2)}{\boldsymbol{k}\Delta/2} = \frac{\sin(\pi\boldsymbol{k}/2\boldsymbol{k}_c)}{\pi\boldsymbol{k}/2\boldsymbol{k}_c} \tag{2.32}$$

A Gaussian filter, a Gaussian distribution with a zero mean and a dispersion of $\Delta^2/12$, is written as

$$\boldsymbol{G}(\boldsymbol{x}) = \left(\frac{6}{\pi\Delta^2}\right)^{1/2} \exp\left(-\frac{6|\boldsymbol{x}|^2}{\Delta^2}\right) \tag{2.33}$$

and its transfer function is

$$\tilde{\boldsymbol{G}}(\boldsymbol{k}) = \exp\left(-\frac{(\Delta\boldsymbol{k})^2}{24}\right) = \exp\left(-\frac{\pi^2}{24}\left(\frac{\boldsymbol{k}}{\boldsymbol{k}_c}\right)^2\right) \tag{2.34}$$

A spectrum filter

$$\boldsymbol{G}(\boldsymbol{x}) = \frac{\sin(\pi\boldsymbol{x}/\Delta)}{\pi\boldsymbol{x}} \tag{2.35}$$

works as a sharp low-pass filter in the spectrum space

$$\tilde{\boldsymbol{G}}(\boldsymbol{k}) = \boldsymbol{H}(\boldsymbol{k}_c - |\boldsymbol{k}|) \tag{2.36}$$

These filter profiles are illustrated in Fig. 2.4.

Invariant properties of the Navier–Stokes equation for translated and rotated frames of reference need to be preserved after the filtering operation (see details in Speziale[23]). Thus, the invariance should be examined, especially when a highly deformed grid system is introduced in the computation; that is, the results could depend on such coordinates containing different cut-off lengths Δ, which would change the wave-number range of the SG energy to be modeled.

2.3.3. *Leonard's decomposition*

The filtered Navier–Stokes equation is written as

$$\frac{\partial \overline{u_i}}{\partial t} + \frac{\partial}{\partial x_j}\overline{u_i u_j} = -\frac{\partial \overline{p}}{\partial x_i} + \nu\frac{\partial}{\partial x_j}\left(\frac{\partial \overline{u_i}}{\partial x_j} + \frac{\partial \overline{u_j}}{\partial x_i}\right) \tag{2.37}$$

Using the relation in Eq. (2.27), the filtered nonlinear term on the left hand side of Eq. (2.37) is decomposed as

$$\overline{u_i u_j} = \overline{(\overline{u_i} + u_i')(\overline{u_j} + u_j')} = \overline{\overline{u_i}\,\overline{u_j}} + \overline{\overline{u_i} u_j'} + \overline{\overline{u_j} u_i'} + \overline{u_i' u_j'} \tag{2.38}$$

It should be noted that cross-stress terms ($\overline{\overline{u_i}u'_j}$, $\overline{\overline{u_j}u'_i}$) generally cannot be eliminated, different from the Reynolds decomposition (see Eq. (2.28) and Fig. 2.2). Substituting Eq. (2.38) into (2.37), we have

$$\frac{\partial \overline{u_i}}{\partial t}+\frac{\partial}{\partial x_j}\overline{\overline{u_i}\,\overline{u_j}}=-\frac{\partial \overline{p}}{\partial x_i}+\nu\frac{\partial}{\partial x_j}\left(\frac{\partial \overline{u_i}}{\partial x_j}+\frac{\partial \overline{u_j}}{\partial x_i}\right)-\frac{\partial}{\partial x_j}\left(\overline{\overline{u_i}u'_j}+\overline{\overline{u_j}u'_i}+\overline{u'_iu'_j}\right) \tag{2.39}$$

The SG stress tensor τ_{ij} provides the contributions of small-scale velocity fluctuations to the larger GS motion and is defined to be

$$\tau_{ij}=C_{ij}+R_{ij}=\overline{u_iu_j}-\overline{\overline{u_i}\,\overline{u_j}} \tag{2.40}$$

where C_{ij} is the cross-stress tensor ($=\overline{\overline{u_i}u'_j}+\overline{\overline{u_j}u'_i}$) and R_{ij} represents the Reynolds SG tensor($=\overline{u'_iu'_j}$). Eq. (2.39) is now described as

$$\frac{\partial \overline{u_i}}{\partial t}+\frac{\partial}{\partial x_j}\overline{\overline{u_i}\,\overline{u_j}}=-\frac{\partial \overline{p}}{\partial x_i}+\nu\frac{\partial}{\partial x_j}\left(\frac{\partial \overline{u_i}}{\partial x_j}+\frac{\partial \overline{u_j}}{\partial x_i}\right)-\frac{\partial \tau_{ij}}{\partial x_j} \tag{2.41}$$

The filtered form of the nonlinear GS velocities, $\overline{\overline{u_i}\,\overline{u_j}}$, are still uncomputable. To decompose this term, a Leonard stress is defined:

$$L_{ij}=\overline{\overline{u_i}\,\overline{u_j}}-\overline{u_i}\,\overline{u_j} \tag{2.42}$$

Eq. (2.41) becomes

$$\frac{\partial \overline{u_i}}{\partial t}+\frac{\partial}{\partial x_j}\overline{u_i}\,\overline{u_j}=-\frac{\partial \overline{p}}{\partial x_i}+\nu\frac{\partial}{\partial x_j}\left(\frac{\partial \overline{u_i}}{\partial x_j}+\frac{\partial \overline{u_j}}{\partial x_i}\right)-\frac{\partial \tau_{ij}}{\partial x_j} \tag{2.43}$$

where τ_{ij} has been redefined to become

$$\tau_{ij}=L_{ij}+C_{ij}+R_{ij}=\overline{u_iu_j}-\overline{u_i}\,\overline{u_j} \tag{2.44}$$

The derived form of the filtered Navier–Stokes equation Eq. (2.43) is analogous to the Reynolds equation Eq. (2.6); thus the identical closure procedure is also required when modeling the SG stress Eq. (2.44).

The kinetic energy of small-scale fluid motion provides a turbulent energy spectrum in wave numbers higher than k_c; this would be an essential parameter for the closure model, which is followed by a SG kinetic energy transport equation derived below.

Multiplying u_i with the Navier–Stokes equation and filtering, we get

$$\frac{1}{2}\frac{\partial \overline{u_iu_i}}{\partial t}+\frac{\partial}{\partial x_j}\overline{u_iu_iu_j}-\overline{u_iu_j\frac{\partial u_i}{\partial x_j}}=-\frac{\partial \overline{u_ip}}{\partial x_i}+\nu\frac{\partial^2\overline{u_iu_j}}{\partial x_j\partial x_j}-\nu\overline{\frac{\partial u_j}{\partial x_i}\frac{\partial u_i}{\partial x_j}} \tag{2.45}$$

Eq. (2.43) is multiplied by $\overline{u_i}$ and expressed as

$$\frac{1}{2}\frac{\partial \overline{u_i}\,\overline{u_i}}{\partial t} + \frac{\partial}{\partial x_j}\overline{u_i}\,\overline{u_i}\,\overline{u_j} - \overline{u_i}\,\overline{u_j}\frac{\partial \overline{u_i}}{\partial x_j}$$
$$= -\frac{\partial \overline{u_i}\,\overline{p}}{\partial x_i} + \nu\frac{\partial^2 \overline{u_i}\,\overline{u_i}}{\partial x_j \partial x_j} - \nu\frac{\partial \overline{u_i}}{\partial x_j}\frac{\partial \overline{u_j}}{\partial x_i} - \frac{\partial}{\partial x_j}\overline{u_i}\tau_{ij} + \tau_{ij}\frac{\partial \overline{u_i}}{\partial x_j} \tag{2.46}$$

Subtracting Eq. (2.46) from Eq. (2.45) and then applying a transport equation for the SG kinetic energy $k_T = \frac{1}{2}(\overline{u_i u_i} - \overline{u_i}\,\overline{u_i}) \equiv \frac{1}{2}\overline{u_i' u_i'}$ yields

$$\frac{\partial k_T}{\partial t} + \frac{\partial}{\partial x_j}k_T\overline{u_j} = -\frac{1}{2}\left(\overline{u_i u_i u_j} - \overline{u_j}\,\overline{u_i u_i}\right) - \frac{\partial}{\partial x_j}\left(\overline{p u_j} - \overline{p}\,\overline{u_j}\right) \tag{2.47}$$
$$+ \frac{\partial}{\partial x_j}\left(\nu\frac{\partial k_T}{\partial x_j}\right) - \nu\left(\overline{\frac{\partial u_i}{\partial x_j}\frac{\partial u_i}{\partial x_j}} - \frac{\partial \overline{u_i}}{\partial x_j}\frac{\partial \overline{u_i}}{\partial x_j}\right) + \frac{\partial}{\partial x_j}\overline{u_i}\tau_{ij} - \tau_{ij}\frac{\partial \overline{u_i}}{\partial x_j}$$

The left hand side of the above equation represents advection of k_T, and terms one through six of the right hand side indicate mechanical contributions of turbulent transport, diffusion due to pressure, viscous diffusion, viscous dissipation, SG diffusion and SG dissipation, respectively. In LES, to determine the SG stress τ_{ij} that is required to compute Eq. (2.43), first solve Eq. (2.47) coupled with a suitable closure model for the unresolved first, second and third terms to determine k_T.

2.3.4. *Smagorinsky model*

Assuming that the process of transferring energy to smaller scales via a cascade-down process is analogous to that of molecular diffusion, the SG stress may be explicitly modeled with an analogy to the eddy viscosity model Eq. (2.9) provided in RANS:

$$-\frac{\partial \tau_{ij}^d}{\partial x_j} = \frac{\partial}{\partial x_j}\nu_T\left(\frac{\partial \overline{u_i}}{\partial x_j} + \frac{\partial \overline{u_j}}{\partial x_i}\right) \tag{2.48}$$

where ν_T is the SG viscosity, and τ_{ij}^d is the deviation tensor with respect to τ_{ij}, which is defined as

$$\tau_{ij}^d \equiv \tau_{ij} - \frac{1}{3}\tau_{kk}\delta_{ij} \tag{2.49}$$

The symmetric components $\frac{1}{3}\tau_{kk}\delta_{ij}$ can be included in the symmetric pressure terms, and therefore modeling is not required for these terms.

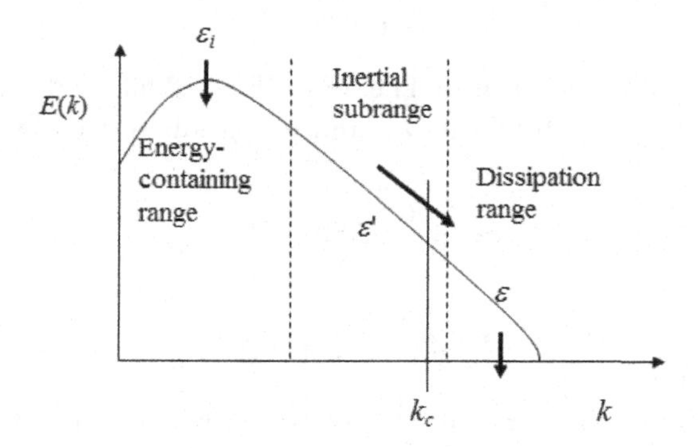

Fig. 2.5. Schematic representation of the local equilibrium assumption for an energy spectrum ε_i(energy production) $= \varepsilon'$(energy flux passing through cut-off wave-number) $= \varepsilon$(viscous dissipation).

The filtered momentum equation for the SG viscosity model is finally written as

$$\frac{\partial \overline{u_i}}{\partial t} + \frac{\partial}{\partial x_j}\overline{u_i}\ \overline{u_j} = -\frac{\partial \overline{P}}{\partial x_i} + \nu\frac{\partial}{\partial x_j}\left(\frac{\partial \overline{u_i}}{\partial x_j} + \frac{\partial \overline{u_j}}{\partial x_i}\right) + \frac{\partial}{\partial x_j}\nu_T\left(\frac{\partial \overline{u_i}}{\partial x_j} + \frac{\partial \overline{u_j}}{\partial x_i}\right) \tag{2.50}$$

where $\overline{P} = \overline{p} + \frac{1}{3}\tau_{kk}$.

Assuming that the energy spectrum for an isotropic homogeneous turbulent flow is in local equilibrium and that there is no temporal change in the spectrum form, the energy production (ε_i), dissipation (ε) and the energy flux (ε') passing through the cut-off wave number k_c are identically given (see Fig. 2.5):

$$\varepsilon_i = \varepsilon' = \varepsilon \tag{2.51}$$

The analogy between the molecular viscous dissipation and the energy flux due to the SG viscosity leads to

$$\varepsilon = \varepsilon' \equiv \tau_{ij}^d \overline{S_{ij}} = 2\nu_T \overline{S_{ij}}\ \overline{S_{ij}} \tag{2.52}$$

where the resolved strain rate $\overline{S_{ij}} = \frac{1}{2}\left(\partial\overline{u_i}/\partial x_j + \partial\overline{u_j}/\partial x_i\right)$. The energy transfer from GS to SG scales is assumed to be governed by the kinetic energy in the vicinity of the cut-off wave number. The Smagorinsky viscosity ν_T and dissipation ε may therefore be provided through dimensional scaling with the SG kinetic energy k_T and cut-off length Δ by

$$\nu_T = C_\nu k_T^{1/2}\Delta \tag{2.53}$$

$$\varepsilon = -C_\varepsilon k_T^{3/2}/\Delta \tag{2.54}$$

where C_ν and C_ε are constants. The above Eqs. (2.53) and (2.54) are substituted into Eq. (2.52) to obtain k_T and ν_T parameterized by the resolved strain:

$$k_T = \frac{C_\nu}{C_\varepsilon}\Delta^2 2\overline{S_{ij}}\ \overline{S_{ij}} \tag{2.55}$$

$$\nu_T = \sqrt{\frac{C_\nu}{C_\varepsilon}}C_\nu\Delta^2(2\overline{S_{ij}}\ \overline{S_{ij}})^{1/2} = (C_s\Delta)^2(2\overline{S_{ij}}\ \overline{S_{ij}})^{1/2} \tag{2.56}$$

where the Smagorinsky constant C_s is explicitly determined by the local equilibrium assumption as described below. Eqs. (2.56) and (2.52) lead to

$$\varepsilon' = 2(C_s\Delta)^2(2\overline{S_{ij}}\ \overline{S_{ij}})^{1/2}\ \overline{S_{ij}}^2 \tag{2.57}$$

On the other hand, the universal form of the Kolmogorov spectrum for isotropic turbulence is given by

$$E(k) = C\varepsilon^{2/3}k^{-5/3} \tag{2.58}$$

where the Kolmogorov constant $C \sim 1.4$. The spectrum representation of the absolute strain rate $\overline{S_{ij}}^2$ is expressed as

$$|\overline{S}|^2 = \overline{S_{ij}}\ \overline{S_{ij}} = \int_0^{\pi/\Delta} k^2 E(k)dk \tag{2.59}$$

Eq. (2.58) is substituted into the above equation, and after integrating we have

$$|\overline{S}|^2 = \pi^{4/3}C\frac{3}{4}\varepsilon^{2/3}\Delta^{-4/3} \tag{2.60}$$

Thus we find ε is expressed by

$$\varepsilon = \left(\frac{1}{\pi}\right)^2\left(\frac{3C}{2}\right)^{-3/2}\Delta^2 2|\overline{S_{ij}}|^{3/2} \tag{2.61}$$

With Eqs. (2.51), (2.57) and (2.61), C_S is determined as

$$C_s = \frac{1}{\pi}\left(\frac{3C}{2}\right)^{-3/4} \sim 0.18 \tag{2.62}$$

In a computational procedure, the Smagorinsky viscosity Eq. (2.56) and its derived constant Eq. (2.62) are computed with the resolved strain rate, which is the only parameter to determine the SG stress Eq. (2.48):

$$-\tau_{ij}^d = 2(C_s\Delta)^2\left(2\overline{S_{ij}}\ \overline{S_{ij}})^{1/2}\overline{S_{ij}}\right) \tag{2.63}$$

Since this is the simplest SG model that explicitly provides the SG stress without any additional equation to solve, this model has been applied to numerous applications across a vast field of research. The user needs to remember that this model was derived based on a fundamental premise, that there is a local equilibrium of energy transfer. This premise requires careful examination of the computed results through experimental comparisons; significant discrepancies between computed and experimentally determined flows have been reported in many cases (see details in Meneveau and Katz[18]). Also, it is difficult to estimate the optimal Smagorinsky constant for arbitrary flows, and other empirical values of C_s in the range of 0.1 to 0.2 estimated to be consistent with DNS and experiments have been used.

Given the assumption of spectrum equilibrium, this model should not be adopted to transitional turbulence where the spectrum form varies in time and space. For instance, in the surf zone large-scale vortices form by an overturning wave, and the smaller turbulence generated during a splash-up cycle is a typical example of unsteady, inhomogeneous turbulence, which obviously violates the model assumption. Christensen[5] compared wave-breaking flows computed with the Smagorinsky and one-equation model and quantified the sensitivities of the turbulence statistics models applied.

Another requirement of this model stems from the assumption of Eq. (2.51) for a single fixed value of k_c: sufficiently fine grids must be introduced to resolve the velocity fluctuations in the inertial subrange of the Kolmogorov spectrum of Eq. (2.58). Also, local energy transfer in a local turbulent structure may not be reasonably computed since a single global constant is used for the entire state of turbulence. The model and coordinate properties should be examined when the model is applied to coastal flows that typically involve unsteady turbulence under waves.

A number of other models have been developed to avoid these problems. A one-equation model to estimate the local SG viscosity from the SG kinetic energy transport equation of Eq. (2.47) is introduced in §2.3.5, and a dynamic model to determine the local Smagorinsky constants at each time step is explained in §2.3.6.

2.3.5. *One-equation model*

The SG kinetic energy for the Kolmogorov spectrum is defined to be

$$k_T \equiv \frac{1}{2}\overline{u_i' u_i'} = \int_{k_c}^{\infty} E(k)dk = \frac{3}{2}C\varepsilon^{2/3}k_c^{-2/3} \tag{2.64}$$

This leads to a dissipation

$$\varepsilon = \frac{k_c}{(3C/2)^{3/2}} k_T^{3/2} \tag{2.65}$$

The above equation is substituted into the energy flux equation Eq. (2.54) through the cut-off wave number k_c to obtain the constants $C_\varepsilon = \pi(3C/2)^{-3/2} \sim 1$ and $C_\nu = \pi^{-1}(3C/2)^{-3/2} \sim 0.1$. The SG viscosity is now explicitly given using C_ν with Eq. (2.53). Note that these coefficients values may depend on statistical assumptions. For instance, in a two-scale direct interference approximation (Yoshizawa and Horiuti[30]), $C_\nu \sim 0.043$ and $C_\varepsilon \sim 1.8$. These values give a Smagorinsky constant $C_s \sim 0.081$, which is very different from the value determined by Eq. (2.62).

Here, we follow the manner of Yoshizawa and Horiuti[30] to model the SG kinetic energy equation Eq. (2.47) and their constants. According to the Kolmogorov-Prandtl relation, the diffusion term of Eq. (2.47) is approximated as

$$\begin{aligned} \frac{\partial}{\partial x_j}(\overline{u_i u_i u_j} - \overline{u_i u_i}\ \overline{u_j}) + \frac{\partial}{\partial x_j}(\overline{u_i p} - \overline{u_i}\ \overline{p}) &\equiv \frac{\partial}{\partial x_j}\left(\frac{1}{2}\overline{u_i' u_i' u_j'} + \overline{u_j' p}\right) \\ &= \frac{\partial}{\partial x_j}\left(C_{kk}\Delta\sqrt{k_T}\frac{\partial k_T}{\partial x_j}\right) \end{aligned} \tag{2.66}$$

where $C_{kk} \sim 0.11$. The dissipation term is modeled as

$$\nu\left(\overline{\frac{\partial u_i}{\partial x_j}\frac{\partial u_i}{\partial x_j}} - \overline{\frac{\partial u_i}{\partial x_j}}\ \overline{\frac{\partial u_i}{\partial x_j}}\right) = \nu\overline{\frac{\partial u_i'}{\partial x_j}\frac{\partial u_i'}{\partial x_j}} = C_\varepsilon k_T^{3/2}/\Delta \tag{2.67}$$

where $C_\varepsilon \sim 1.8$. Thus the modeled SG kinetic transport equation becomes

$$\begin{aligned} \frac{\partial k_T}{\partial t} + \frac{\partial}{\partial x_j}k_T\overline{u_j} = -\tau_{ij}\left(\frac{\partial \overline{u_i}}{\partial x_j} + \frac{\partial \overline{u_j}}{\partial x_i}\right) - C_\varepsilon k_T^{3/2}/\Delta \\ + C_{kk}\frac{\partial}{\partial x_j}\Delta\sqrt{k_T}\frac{\partial k_T}{\partial x_j} + \nu\frac{\partial^2 k_T}{\partial x_j \partial x_j} \end{aligned} \tag{2.68}$$

The local SG stress is determined by substituting the computed k_T into Eq. (2.53). This model can account for the local energy transfer in the subgrid range of the non-equilibrium spectrum and theoretically provides a better solution than the Smagorinsky model.

2.3.6. *Dynamic model*

As already mentioned, the Smagorinsky model is described by the unique global constant for the equilibrium spectrum, which cannot describe transitional turbulence and local coherent structures. Germano[8] proposed a dynamic model through double filtering, where local constants are adapted to local flows and are reasonably estimated for the entire computing grid at every time step.

Consider the filtering operations of a fine filter (F-level filtering) and coarse filter (C-level filtering) with different cut-off lengths, one shorter and one longer. The SG tensor for both the F and C operations ($\tau_{FC}(u_i, u_j) = \widetilde{\overline{u_i u_j}} - \widetilde{\overline{u_i}}\,\widetilde{\overline{u_i}}$) is defined to be the Germano identity:

$$\tau_{FC}(u_i, u_j) = \widetilde{\tau_F}(u_i, u_j) + \tau_F(\overline{u_i}, \overline{u_j}) \tag{2.69}$$

where $\widetilde{\tau_C}(u_i, u_j) = \widetilde{\overline{u_i u_j}} - \widetilde{\overline{u_i}\,\overline{u_i}}$ and $\tau_F(\overline{u_i}, \overline{u_j}) = \widetilde{\overline{u_i}\,\overline{u_j}} - \widetilde{\overline{u_i}}\,\widetilde{\overline{u_i}}$.

The cut-off length of the fine level filter F is typically identical with the grid spacing $\overline{\Delta}$ in physical space, and in the coarser one C (called a test filter) typically uses a longer scale $\widetilde{\overline{\Delta}} = 2\overline{\Delta}$ of twice the grid width.

The Germano identity of Eq. (2.69) follows the relation

$$L_{ij} = \widetilde{\overline{u_i}\,\overline{u_j}} - \widetilde{\overline{u_i}}\,\widetilde{\overline{u_j}} = T_{ij} - \widetilde{\tau_{ij}} \tag{2.70}$$

where

$$\tau_{ij} = \overline{u_i u_j} - \overline{u_i}\,\overline{u_j} \tag{2.71}$$

and

$$T_{ij} = \widetilde{\overline{u_i u_j}} - \widetilde{\overline{u_i}}\,\widetilde{\overline{u_j}} \tag{2.72}$$

Assuming the τ_{ij} and T_{ij} for both of the filtering levels are locally described by the identical constant C_d, these SG stresses on both scales may be modeled by the Smagorinsky form of Eq. (2.63):

$$\tau_{ij} - \frac{1}{3}\tau_{kk}\delta_{ij} = C_d\beta_{ij} \tag{2.73}$$

$$T_{ij} - \frac{1}{3}T_{kk}\delta_{ij} = C_d\alpha_{ij} \tag{2.74}$$

where the deviation tensors β_{ij} and α_{ij} are determined by

$$\beta_{ij} = -2\overline{\Delta}^2|\overline{S}|\overline{S_{ij}} \tag{2.75}$$

$$\alpha_{ij} = -2\widetilde{\overline{\Delta}}^2|\widetilde{\overline{S}}|\widetilde{\overline{S_{ij}}} \tag{2.76}$$

The above relations are substituted into Eq. (2.70) to yield

$$L_{ij} - \frac{1}{3}L_{kk}\delta_{ij} \equiv L_{ij}^{d} = C_d\alpha_{ij} - \widetilde{C_d\beta_{ij}} \tag{2.77}$$

Using the approximation $\widetilde{C_d\beta_{ij}} \equiv C_d\widetilde{\beta_{ij}}$, then the residual of Eq. (2.77), E_{ij} is given as

$$E_{ij} = L_{ij} - \frac{1}{3}L_{kk}\delta_{ij} - C_d\alpha_{ij} + C_d\widetilde{\beta_{ij}} \tag{2.78}$$

Optimal values of C_d can be determined with the least squares method (Lilly[14]); that is, the optimal condition to minimize the residual

$$\frac{\partial E_{ij}E_{ij}}{\partial C_d} = 0 \tag{2.79}$$

leads to

$$C_d = \frac{m_{ij}L_{ij}^{d}}{m_{kl}m_{kl}} \tag{2.80}$$

where $m_{ij} = \alpha_{ij} - \widetilde{\beta_{ij}}$. Eqs. (2.73) and (2.75) coupled with the optimal C_d value dynamically provide the local SG stress, which gives a better description of the non-equilibrium state of turbulence.

2.4. Applications of Turbulence Models in the Surf Zone

RANS and LES are widely used to compute turbulent flows in diverse fields of research including coastal and ocean engineering. In the RANS, since any fluid motion with an arbitrary turbulence scale is simply modeled as a Reynolds stress (*i.e.*, an explicit scale separation is not performed), the computing grids may be specifically designed to resolve fluctuations of ensemble mean flows. That is, grids coarser than typically used for LES and DNS can be applicable, and lower computational expenses can be expected. This allows for a wide range of model applications, from local flow at a small experimental scale (*e.g.* turbulent boundary layer flows in a channel) to global ocean current models.

As found from Eq. (2.3), the RANS model defines only random turbulence fluctuations described by the Kolmogorov spectrum. Therefore this premise must be satisfied; otherwise erroneous excessive diffusion may be induced, especially for statistically unsteady flows in a state of non-equilibrium, or organized fluctuations that are not governed by the assumed turbulence statistics but rather large-scale nonlinear interactions.

Unsteady Reynolds Averaged Numerical Simulations (URANS) may support these flow components. In this model, the Reynolds decomposition takes the following form instead of Eq. (2.1):

$$u_i = U_i + \langle u_i \rangle + u_i' \tag{2.81}$$

where U_i and u_i' are the statistical averaged velocity and deviation velocity component, respectively. The term $\langle u_i \rangle$ is the statistically unsteady velocity component with moderate scale variations.

Taking the Reynolds average of the momentum equation with the above decomposition, additional uncomputable terms that represent the correlations for $\langle u_i \rangle$ are produced in the governing equation. Therefore, another closure procedure is required to model these terms. The details of this model can be found in Bosch and Rodi.[2]

In general, a DNS for turbulent flows requires very expensive computations using very fine grids to resolve the entire range of velocity fluctuations. As such, the computable flow condition and size of the domain may be limited. While these limitations can be relaxed in LES, the cut-off wavenumber k_c needs to be in a wavenumber range higher than the inertial subrange, especially for the Smagorinsky model where the energy spectrum is assumed to be in a local equilibrium. Thus, the computing grids need to be fine enough and comparable to a Kolmogorov length scale, which defines the wave-number for the lower limit of the dissipation range (see Fig. 2.5). The Kolmogorov length is defined as

$$\eta = \left(\nu^3/\varepsilon\right)^{1/4} \tag{2.82}$$

where ν and ε are the dynamic viscosity and dissipation, respectively. For flow in the surf zone on a laboratory experimental scale, η can be estimated to be $O(1\mu m)$ if the dissipation due to wave breaking is assumed to be $\varepsilon = 10W/kg$ (with the dynamic viscosity of water $\nu = 10^{-6}m^2/s$). Whereas grids several times larger than η may be used in practical computations, this resolution is still very fine when covering the entire surf zone, and the computational costs will be expensive. Therefore, in terms of computational ability, only laboratory scales of computation may be realistic at this time. Further improvements in computers are expected to enable the computation of *in-situ* LES on a large oceanic domain.

Both RANS and LES have been applied to turbulent flows in the surf zone, as described below. Lin and Liu[15] performed RANS computations of spilling waves propagating over a uniform 1/35 slope in a two-dimensional (horizontal-vertical plane) wave flume. A VOF method (described in Chapter 4) to capture the free-surface and the standard $k-\varepsilon$ turbulence model were used in the computations. They compared the time records of the

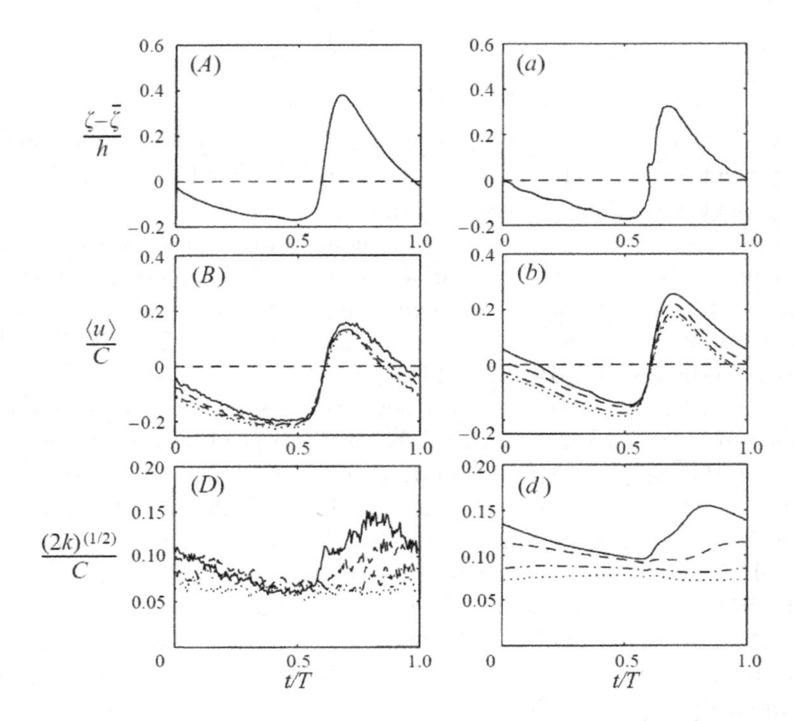

Fig. 2.6. Time records of surface elevation for a spilling breaker (A, a), horizontal velocity (B, b) and turbulent energy (D, d). The computed results by RANS[15] are on the right and experimental ones[26] are on the left (ζ: surface elevation, h: water depth, C: wave celerity).

experimental results[26] with the computed surface elevation, horizontal velocity, and turbulent energy at a fixed point located just outside of the surf zone, as shown in Fig. 2.6.

While overall features of the modeled parameters at this location were consistent with experimental ones, they found that the computed turbulent energy was overestimated near the wave breaking point since the $k-\varepsilon$ model may not reasonably estimate anisotropic turbulence produced in strong shears that appear at breaking wave crests.[15] After wave breaking, the local turbulent motion evolving at the steep crest develops into a turbulent bore containing fully-developed turbulence during the wave breaking process. The RANS computation in this bore region was found to reasonably model these turbulent properties. They also found that the computed result is very sensitive to the model constant $C_{\varepsilon 1}$ in the transport equation for ε,

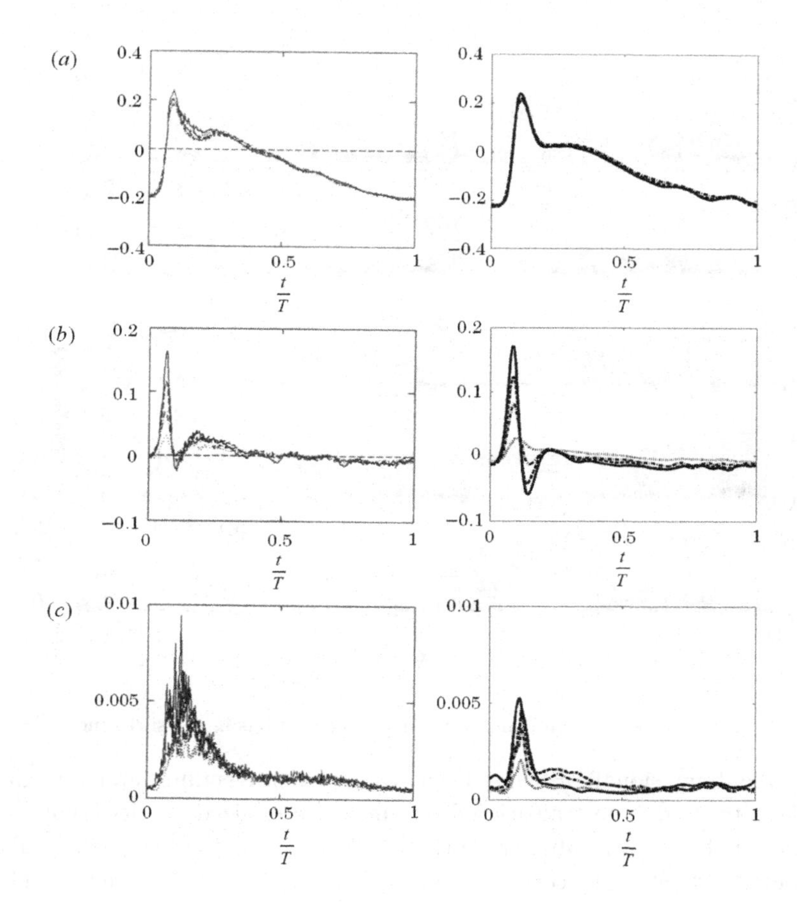

Fig. 2.7. Time records of phase-averaged horizontal velocity for a plunging breaker: (a) $\langle\overline{u}\rangle/C$, (b) vertical velocity $\langle\overline{w}\rangle/C$, and (c) turbulent energy k_T/C^2. Computed results by LES[28] (right) and experimental ones[25] (left) (C: wave celerity).

Eq. (2.19). For instance, a 10% change in $C_{\varepsilon 1}$ produces a 50% change in k, which also affects the free-surface elevation and the velocity field.

As previously explained, since the constants used in the standard $k-\varepsilon$ model were empirically determined for steady sheared turbulence,[13] there is no guarantee that the same constants are also applicable to anisotropic, unsteady, organized turbulence that is typically formed during wave breaking. Further study on optimal constants and higher order models is expected to extend computational breaking-wave research.

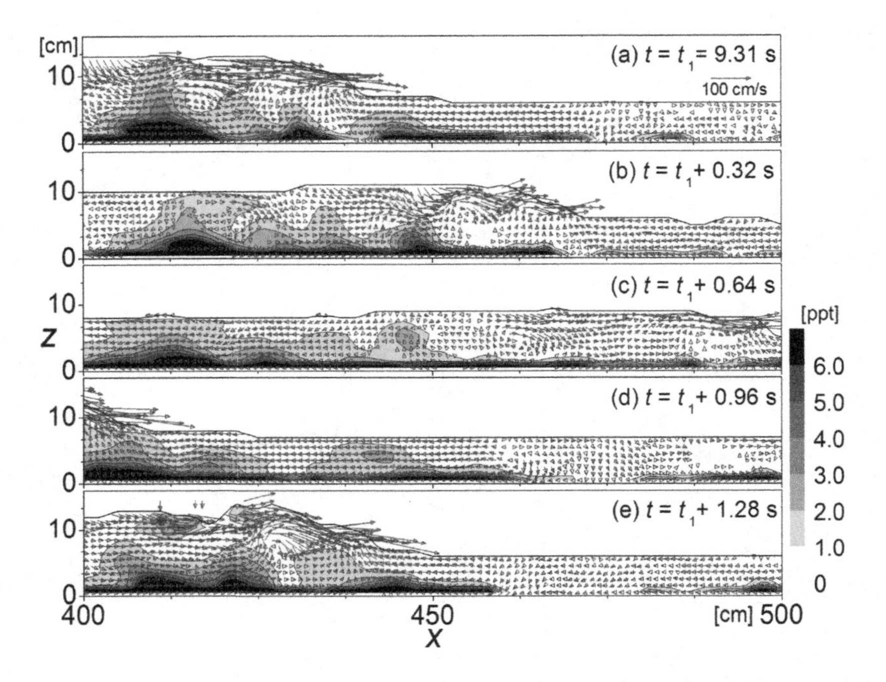

Fig. 2.8. Velocity vectors and sediment concentrations in the surf zone.[24]

Three-dimensional LES has been performed for spilling and plunging breakers to study the transitional features of organized vortices and turbulence that are both inherent in the breaking processes[28] (Fig. 2.7). The computed velocity and turbulent energy values in the surf zone are also compared with the laboratory experiments done by Ting and Kirby.[25] The temporal evolutions of the computed horizontal and vertical velocities that appear in the energetic splashing process are consistent with the experimental ones. While the maximum turbulent energy is underestimated, this may be caused by a single phase assumption that does not incorporate aeration in the computation. That is, various sizes of air bubbles are entrained in the surf zone, which may intensify the fluid turbulence around the bubbles[1] or suppress it, depending on the relative size of the bubble and the turbulence length-scale.[9] An appropriate aeration model that accounts for the entrainment of unresolved bubbles and the mechanical interactions with turbulence is necessary for future coastal research. It should be noted that, despite applaudable efforts with post-processing of discontinuous velocity signals, physically measuring turbulence in aerated flows is also challenging. This suggests that a quantitative comparison with measured turbulence could involve a certain level of error.

Cox and Kobayashi[6] observed an intermittent feature of turbulence in the surf zone. Since intermittent turbulence also disturbs sediments intermittently, instantaneous turbulence and bottom shear should be essential parameters to estimate the suspensioeen process of bottom sediments. For this type of problem, LES has an advantage over RANS since instantaneous flow can be directly computed; LES applied to sediment transport is shown in Fig. 2.8.[24] The concentration of suspended sediment has been estimated with the computed instantaneous near-bottom velocities and accelerations, resulting in the intermittent and locally intensified sediment suspension previously observed.

References

1. Bunner B. and Tryggvason G. (2002): Dynamics of homogeneous bubbly flows. Part 2. Velocity fluctuations, *Journal of Fluid Mechanics.* Vol.466, pp.53-84.
2. Bosch G. and Rodi W. (1998): Simulation of vertex shedding past a square cylinder with different turbulence models, *International Journal for Numerical Methods in Fluids*, Vol.28, pp.601-616.
3. Chou P.Y. (1945): On velocity correlations and the solutions of the equations of turbulent fluctuation, *Quart Appl. Math*, Vol.3, pp.38-54.
4. Christensen E.D. and Deigaard R. (2001): Large eddy simulation of breaking waves, *Coastal Engineering*, Vol.42, pp.53-86.
5. Christensen E.D. (2006): Large eddy simulation of spilling and plunging breakers, *Coastal Engineering*, Vol.53, pp.463-485.
6. Cox D.T. and Kobayashi N. (2000): Identification of intense, intermittent coherent motions under shoreline and breaking waves, *J. Geophys. Res.*, Vol.105 (C6), pp.14223-14236.
7. Ferziger J.H. and Peric M. (2001): *Computational Methods for Fluid Dynamics*, Springer.
8. Germano M., Piomelli U., Moin P. and Cabot W.H. (1991): A dynamic subgrid-scale eddy viscosity model, *Physics of Fluids*, A, Vol.3, pp.1760-1765.
9. Gore R.A. and Crowe C.T. (1989): Effect of particle size on modulating turbulent intensity, *Int. J. Multiphase Flow*, Vol.15, pp.279-285.
10. Jones W.P. and Launder B.E. (1972): The prediction of laminarization with two equation model of turbulence, *International Journal of Heat and Mass Transfer*, Vol.15, pp.301-314.
11. Kolmogorov A. (1941): The local structure of turbulence in incompressible viscous fluid for very large Reynolds numbers, *Dokl. Akad. Nauk SSSR*, Vol.30, pp.301-305.
12. Labourasse E., Lacanette D., Toutant A., Lubin P., Vincent S., Lebaigue O., Caltagirone J.-P. and Sagaut P. (2007): Towards large eddy simulation of

isothermal two-phase flows: Governing equations and a priori tests, *International Journal of Multiphase Flow*, Vo;.33, pp.1-39.

13. Launder B. and Spalding D.B. (1974): The numerical computation of turbulent flows, *Computer Methods in Applied Mechanics and Engineering*, Vol.3, pp.269-289.
14. Lilly D.K. (1992): A proposed modification of the Germano subgrid-scale closure method, *Physics of Fluids*, A 4(3), pp.633-635.
15. Lin P. and Liu P.L.-F. (1998): A numerical study of breaking waves in the surf zone, *Journal of Fluid Mechanics*, Vol.359, pp.239-264.
16. Mansour N.N., Kim J. and Moin P. (1987): Reynolds stress and dissipation rate budgets in turbulent channel flow, *Journal of Fluid Mechanics*, Vol.194, pp.15-44.
17. Mellor G.L. and Yamada T. (1974): A hierarchy of turbulence closure models for planetary boundary layers. Journal of Atmospheric Sciences, Vol.31, pp.1791-1806.
18. Meneveau C. and Katz J. (2000): Scale-invariance and turbulence models for large-eddy simulation, *Annual Review of Fluid Mechanics*, Vol.32, pp.1-32.
19. Piomelli U. (1999): Large-eddy simulation: achievements and challenges, *Progress in Aerospace Sciences*, Vol.35, pp.335-362.
20. Rotta J.C. (1968): Turbulent boundary layers in incompressible flow, *Progress in Aerospace Sciences*, Vol.2, p.1.
21. Saffman P.G. (1970): A model for inhomogeneous turbulent flow, *Proceedings of Royal Society of London*, Vol.A317, pp.417-433.
22. Sagaut P. (2000): *Large eddy simulation for incompressible flows*, Springer.
23. Speziale C.G. (1985): Galilean invariance of subgrid-scale stress models in the large-eddy simulation of turbulence, *Journal of Fluid Mechanics*, Vol.156, pp.55-62.
24. Suzuki T., Okayasu A. and Shibayama T. (2007): A numerical study of intermittent sediment concentration under breaking waves in the surf zone, *Coastal Engineering*, Vol.54, pp.433-444.
25. Ting F.C.K. and Kirby J.T. (1995): Dynamics of surf-zone turbulence in a strong plunging breaker, *Coastal Engineering*, Vol.24, pp.177-204.
26. Ting F.C.K. and Kirby J.T. (1996): Dynamics of surf-zone turbulence in a spilling breaker, *Coastal Engineering*, Vol.27, pp.131-160.
27. Umlauf L. and Burchard H. (2003): A generic length-scale equation for geophysical turbulence models, *Journal of Marine Research*, Vol.61, No.2, pp. 235-265.
28. Watanabe Y., Saeki H. and Hosking R.J. (2005): Three-dimensional vortex structures breaking waves, *Journal of Fluid Mechanics*, Vol.545, pp.291-328.
29. Wilcox D.C. (1998): *Turbulence Modeling for CFD*, DCW Industries.
30. Yoshizawa A. and Horiuti K. (1985): A statistically-derived subgrid-scale kinetic energy model for the large-eddy simulation of turbulent flows, *Journal of Physical Society of Japan*, Vol.54(8), pp.2834-2839.

Chapter 3

Fundamental Computational Methods

Yasunori Watanabe and Nobuhito Mori

The previous chapter outlined and explained the Navier–Stokes equation. The Navier–Stokes equation describes general fluid flows including the wave field. However, this nonlinear differential equation is very difficult to solve analytically for arbitrary flows. The fluid dynamics of complex flows can be approximated and computed via algebraic solutions of the discretized Navier–Stokes equations using appropriate boundary conditions. Furthermore, a Poisson equation, which is derived from the divergence of the incompressible Navier–Stokes equation, provides the pressure, ensuring a divergence-free velocity field. Due to the assumption of allowing for infinite speed of sound in fluid, this type of differential equation requires a simultaneous solution over the entire computational grid. Because of this, the computational costs become expensive as the total number of grid points in the fluid domain increases. Efficient procedures to solve the simultaneous equations are necessary for practical computations.

In this chapter, fundamental procedures to discretize the differential equations on the basis of the finite difference method are discussed in §3.1. Major boundary conditions used in wave computations are introduced in §3.2. And, practical techniques to solve a Poisson-type differential equation are explained in §3.3.

3.1. Discretization for Finite Differences

Quantities of $f(x_0)$, $f(x_1)$, $f(x_2)$,$\cdots$, $f(x_i)$, $\cdots$, $f(x_N)$ are defined and given at $(1+N)$ discrete coordinates, x_0, x_1, x_2,$\cdots$, x_i, $\cdots$, x_N, respectively. The Taylor series of $f(x_{i+1}) = f(x_i + \Delta_+)$ about x_i is

$$f(x_i + \Delta_+) = f(x_i) + \Delta_+ \frac{\partial f(x_i)}{\partial x} + \frac{\Delta_+^2}{2}\frac{\partial^2 f(x_i)}{\partial x^2} + O(\Delta_+^3) \qquad (3.1)$$

where $\Delta_+ = x_{i+1} - x_i$. The first order approximation of Eq. (3.1) can be written as

$$f(x_i + \Delta_+) = f(x_i) + \Delta_+ \frac{\partial f(x_i)}{\partial x} + O(\Delta_+^2) \tag{3.2}$$

which provides the forward difference with the first-order accuracy

$$\frac{\partial f(x_i)}{\partial x} \simeq \frac{f(x_i + \Delta_+) - f(x_i)}{\Delta_+} \tag{3.3}$$

The Taylor series of $f(x_{i-1}) = f(x_i - \Delta_-)$ about x_i is also expressed by

$$f(x_i - \Delta_-) = f(x_i) - \Delta_- \frac{\partial f(x_i)}{\partial x} + \frac{\Delta_-^2}{2} \frac{\partial^2 f(x_i)}{\partial x^2} + O(\Delta_-^3) \tag{3.4}$$

For the case of uniform grid width $\Delta = \Delta_+ = \Delta_-$, subtracting Eq. (3.4) from Eq. (3.1), we obtain the central difference for the first derivative with second-order accuracy

$$\frac{\partial f(x_i)}{\partial x} \simeq \frac{f(x_i + \Delta) - f(x_i - \Delta)}{2\Delta} \tag{3.5}$$

Adding Eqs. (3.4) and (3.1), the second derivative of f can be approximated by

$$\frac{\partial^2 f(x_i)}{\partial x^2} \simeq \frac{f(x_i + \Delta) - 2f(x_i) + f(x_i - \Delta)}{\Delta^2} \tag{3.6}$$

In nonuniform grids, $\Delta_+ \neq \Delta_-$, second-order approximations of the first and second derivatives can be also derived from the Taylor series Eqs. (3.1) and (3.4):

$$\frac{\partial f(x_i)}{\partial x} \simeq \frac{\Delta_-^2 \left(f(x_i + \Delta_+) - f(x_i)\right) + \Delta_+^2 \left(f(x_i) - f(x_i - \Delta_-)\right)}{\Delta_- \Delta_+ \left(\Delta_- + \Delta_+\right)} \tag{3.7}$$

$$\frac{\partial^2 f(x_i)}{\partial x^2} \simeq 2 \frac{\Delta_- f(x_i + \Delta_+) + \Delta_+ f(x_i + \Delta_-) - \left(\Delta_- + \Delta_+\right) f(x_i)}{\Delta_+ \Delta_- \left(\Delta_+ + \Delta_-\right)} \tag{3.8}$$

The time derivative can be also approximated by the finite difference with a discrete time step assuming a sufficiently short time-step interval Δt:

$$\frac{\partial f(x_i, t)}{\partial t} \simeq \frac{f(x_i, t + \Delta t) - f(x_i, t)}{\Delta t} \tag{3.9}$$

Higher order approximations of the derivatives can be calculated by simple addition and subtraction of the higher order Taylor approximations about multiple distances from x_i to the nearby grid points. For instance, assuming a uniform grid width, the third-order approximations of $f_{i+1} = f(x_i + \Delta)$, $f_{i-1} = f(x_i - \Delta)$ and $f_{i-2} = f(x_i - 2\Delta)$ about x_i are

$$f_{i+1} = f_i + \Delta\frac{\partial f_i}{\partial x} + \frac{\Delta^2}{2}\frac{\partial^2 f_i}{\partial x^2} + \frac{\Delta^3}{6}\frac{\partial^3 f_i}{\partial x^3} + O(\Delta^4) \tag{3.10}$$

$$f_{i-1} = f_i - \Delta\frac{\partial f_i}{\partial x} + \frac{\Delta^2}{2}\frac{\partial^2 f_i}{\partial x^2} - \frac{\Delta^3}{6}\frac{\partial^3 f_i}{\partial x^3} + O(\Delta^4) \tag{3.11}$$

$$f_{i-2} = f_i - 2\Delta\frac{\partial f_i}{\partial x} + 2\Delta^2\frac{\partial^2 f_i}{\partial x^2} - \frac{4\Delta^3}{3}\frac{\partial^3 f_i}{\partial x^3} + O(\Delta^4) \tag{3.12}$$

These provide third-order accuracy for the finite differences of the first, second, and third derivatives at x_i in terms of f_i, f_{i+1}, f_{i-1} and f_{i-2}:

$$\frac{\partial f_i}{\partial x} = \frac{1}{6\Delta}(2f_{i+1} + 3f_i - 6f_{i-1} + f_{i-2}) \tag{3.13}$$

$$\frac{\partial^2 f_i}{\partial x^2} = \frac{1}{\Delta^2}(f_{i+1} - 2f_i + f_{i-1}) \tag{3.14}$$

$$\frac{\partial^3 f_i}{\partial x^3} = \frac{1}{\Delta^3}(f_{i+1} - 3f_i + 3f_{i-1} - f_{i-2}) \tag{3.15}$$

Using the same procedure, the fourth-order approximation of the derivatives can also be calculated in terms of the quantities at four neighboring grid coordinates: x_{i-2}, x_{i-1}, x_{i+1} and x_{i+2}

$$\frac{\partial f_i}{\partial x} = \frac{1}{12\Delta}(-f_{i+2} + 8f_{i+1} - 8f_{i-1} + f_{i-2}) \tag{3.16}$$

$$\frac{\partial^2 f_i}{\partial x^2} = \frac{1}{12\Delta^2}(-f_{i+2} + 16f_{i+1} - 30f_i + 16f_{i-1} - f_{i+2}) \tag{3.17}$$

$$\frac{\partial^3 f_i}{\partial x^3} = \frac{1}{2\Delta^3}(f_{i+2} - 2f_{i+1} + 2f_{i-1} - f_{i-2}) \tag{3.18}$$

$$\frac{\partial^4 f_i}{\partial x^4} = \frac{1}{\Delta^4}(f_{i+2} - 4f_{i+1} + 6f_i - 4f_{i-1} + f_{i-2}) \tag{3.19}$$

In this way, we found that higher-order approximations generally require quantities for many grid points, which is not often feasible near boundaries. Another typical example to show the relation between the accuracy and the number of grids required (support compactness) is explained below.

Assuming a periodic domain $f(x_0) = f(x_n)$ in length L, the Fourier representation of $f(x)$ is

$$f(x_i) = \sum_{n=0}^{N-1} F_n e^{jnkx_i} \tag{3.20}$$

where $k = 2\pi/L$ and j is the imaginary unit. The Fourier coefficient F_n can be computed by Fourier transform of $f(x_n)$ for whole grid points; $F_n =$

$\sum_{i=0}^{N-1} f(x_n)e^{jk_n x_i}$. The derivatives are automatically given by differentiating Eq. (3.20)

$$\frac{\partial f(x_i)}{\partial x} = jk \sum_{n=0}^{N-1} nF_n e^{jnkx_i} \tag{3.21}$$

$$\frac{\partial^2 f(x_i)}{\partial x^2} = -k^2 \sum_{n=0}^{N-1} n^2 F_n e^{jnkx_i} \dots \tag{3.22}$$

$$\frac{\partial^m f(x_i)}{\partial x^m} = (jk)^m \sum_{n=0}^{N-1} n^m F_n e^{jnkx_i} \tag{3.23}$$

The computation for the Fourier representation of the flow field, called the spectrum method, is very accurate and often used in DNS, although a whole set of quantities in the domain is required to determine the derivatives.

Next, classical finite difference techniques to solve differential equations are explained. Consider the following general form of the differential equation

$$\frac{\partial f(x,t)}{\partial t} = g(f) \tag{3.24}$$

According to the trapezoid rule for a differential equation in a discrete system, the variable f can be updated by the following procedure (see Fig. 3.1).

$$f^{n+1} = f^n + \Delta t g^{n+m} \tag{3.25}$$

$$g^{n+m} = (1-m)g^n + mg^{n+1} \tag{3.26}$$

where $g^n = g(f^n)$. For the case where $m = 0$, a forward Euler's rule is

$$\frac{f^{n+1} - f^n}{\Delta t} = g(f^n) \tag{3.27}$$

which is an explicit way to solve the equation. An implicit backward Euler's rule ($m = 1$) is written as

$$\frac{f^{n+1} - f^n}{\Delta t} = g(f^{n+1}) \tag{3.28}$$

For $m = 1/2$, an implicit second-order approximation is given by the Crank–Nicolson method in terms of the mean gradient of the variable (see Fig. 3.1)

$$\frac{f^{n+1} - f^n}{\Delta t} = \frac{1}{2}\left(g(f^n) + g(f^{n+1})\right) \tag{3.29}$$

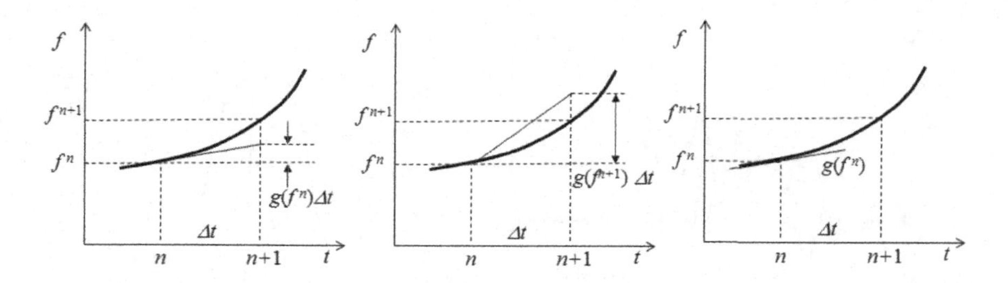

Fig. 3.1. Schematic representation of forward Euler (left, undershooting solution), backward Euler (middle, overshooting solution) and Crank–Nicolson (right) methods.

Based on these procedures, the advection equation is considered solved

$$\frac{\partial f(x,t)}{\partial t} + u\frac{\partial f}{\partial x} = 0 \tag{3.30}$$

Using Eqs. (3.5) and (3.27), Eq. (3.30) can be discretized as

$$\frac{f_i^{n+1} - f_i^n}{\Delta t} + u\frac{f_{i+1}^n - f_{i-1}^n}{2\Delta x} = 0 \tag{3.31}$$

which is called the forward-time central-space (FTCS) method. Substituting the Taylor series of f_i^{n+1}, f_{i+1}^n, and f_{i-1}^n

$$f_i^{n+1} = f_i^n + \Delta t\frac{\partial f_i^n}{\partial t} + \frac{\Delta t^2}{2}\frac{\partial^2 f_i^n}{\partial t^2} + \frac{\Delta t^3}{6}\frac{\partial^3 f_i^n}{\partial t^3} + \cdots \tag{3.32}$$

$$f_{i+1}^n = f_i^n + \Delta x\frac{\partial f_i^n}{\partial x} + \frac{\Delta x^2}{2}\frac{\partial^2 f_i^n}{\partial x^2} + \frac{\Delta x^3}{6}\frac{\partial^3 f_i^n}{\partial x^3} + \cdots \tag{3.33}$$

$$f_{i-1}^n = f_i^n - \Delta x\frac{\partial f_i^n}{\partial x} + \frac{\Delta x^2}{2}\frac{\partial^2 f_i^n}{\partial x^2} - \frac{\Delta x^3}{6}\frac{\partial^3 f_i^n}{\partial x^3} + \cdots \tag{3.34}$$

into the discretized differential Eq. (3.31), we get

$$\frac{\partial f}{\partial t} + u\frac{\partial f}{\partial x} + e(f) = 0 \tag{3.35}$$

where the truncation error is

$$e(f) = cu\frac{\Delta x}{2}\frac{\partial^2 f}{\partial x^2} + u(1 + 2c^2)\frac{\Delta x^2}{6}\frac{\partial^3 f}{\partial x^3} + \cdots \tag{3.36}$$

Here, c is the Courant number defined as $c = u\Delta t/\Delta x$. The order of the truncation error represents the order of accuracy of the numerical scheme. It is obvious that the first term involving the second derivative of f in

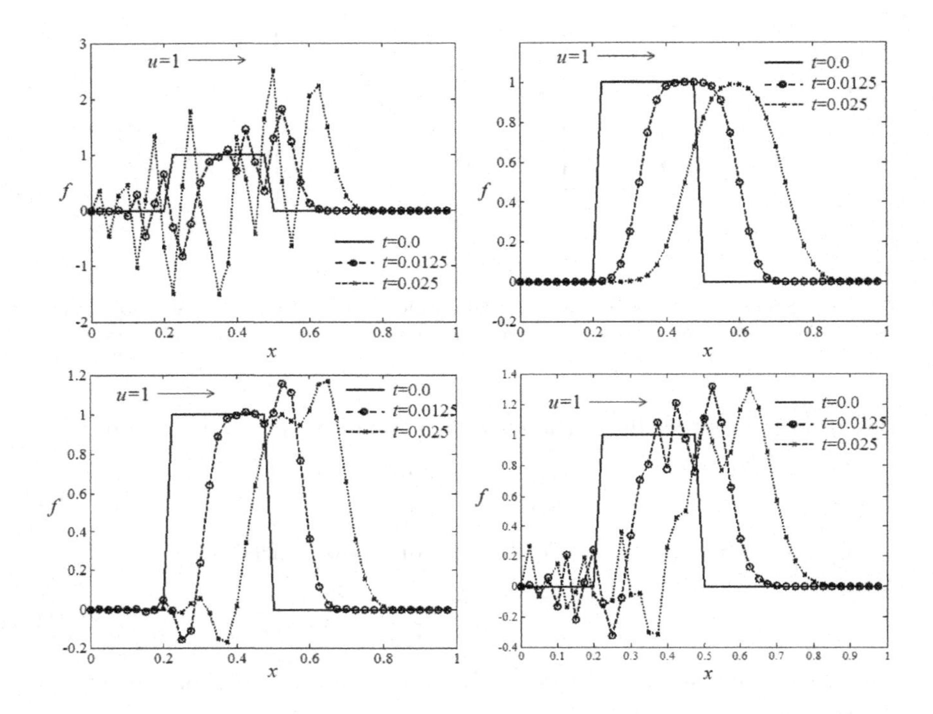

Fig. 3.2. Computational solution of the advection equation ($c = 0.5$) by various methods: FTCS (top left), first-order upwind (top right), Lax–Wendroff (bottom left) and Crank–Nicolson (bottom right).

Eq. (3.36) resembles the negative diffusion term (with a diffusion coefficient of $-cu\Delta x/2$) in the following equation, a term that causes unstable computations and incorrect oscillating solutions (see Fig. 3.2).

$$\frac{\partial f}{\partial t} + u\frac{\partial f}{\partial x} = -cu\frac{\Delta x}{2}\frac{\partial^2 f}{\partial x^2} + O(\Delta x^2) \tag{3.37}$$

This indicates that the computed differential equation may be altered from the original equation we want to solve due to the truncation error that results from the numerical scheme.

The other common, classical schemes and truncation errors shown in Fig. 3.2 are as follows.

First-Order Upwind Method

$$\frac{f_i^{n+1} - f_i^n}{\Delta t} + u\frac{f_i^n - f_{i-1}^n}{\Delta x} = 0 \tag{3.38}$$

$$e(f) = -(1-c)u\frac{\Delta x}{2}\frac{\partial^2 f}{\partial x^2} + u(1 - 3c + 2c^2)\frac{\Delta x^2}{6}\frac{\partial^3 f}{\partial x^3} + \cdots \tag{3.39}$$

This scheme is stable for $c \leq 1$, but the error works as additional unwanted diffusion (numerical viscosity). Thus, a diffusive result may be computed (see Fig. 3.2).

Lax–Wendroff Method

$$\frac{f_i^{n+1} - f_i^n}{\Delta t} + u\frac{f_{i+1}^n - f_{i-1}^n}{2\Delta x} - \frac{1}{2}uc\Delta x\frac{f_{i+1}^n - 2f_i^n + f_{i-1}^n}{\Delta x^2} = 0 \tag{3.40}$$

$$e(f) = u(1-c^2)\frac{\Delta x^2}{6}\frac{\partial^3 f}{\partial x^3} + uc(1-c^2)\frac{\Delta x^3}{8}\frac{\partial^4 f}{\partial x^4} + \cdots \tag{3.41}$$

In this second order scheme, the first term of the truncation error in the FTCS scheme of Eq. (3.36) and has been canceled out with the additional third term of Eq. (3.40).

Crank–Nicolson Method

Based on Eq. (3.29), this method uses the mean gradient of the variable f and yields second-order accuracy.

$$\frac{f_i^{n+1} - f_i^n}{\Delta t} + u\frac{f_{i+1}^n - f_{i-1}^n + f_{i+1}^{n+1} - f_{i-1}^{n+1}}{4\Delta x} = 0 \tag{3.42}$$

$$e(f) = u(1 + c^2/2)\frac{\Delta x^2}{6}\frac{\partial^3 f}{\partial x^3} + \cdots \tag{3.43}$$

We encourage the reader to read additional literature sources that explain the many other numerical schemes in use (e.g. Ferziger *et al.*[1]).

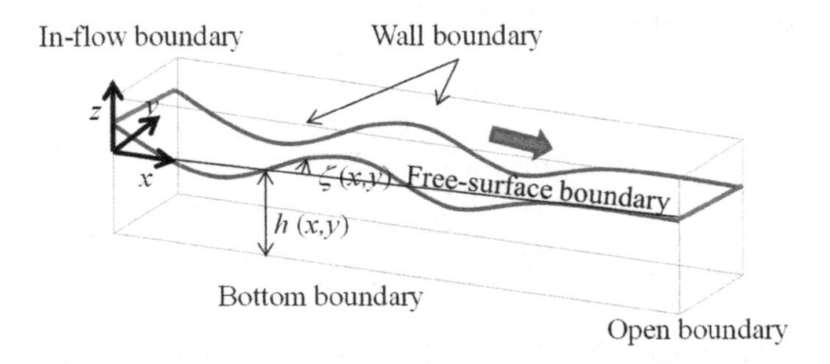

Fig. 3.3. Boundary conditions used in a numerical wave flume.

3.2. Boundary Conditions

At any boundary of the fluid region, certain physical conditions must be incorporated to solve the governing momentum and pressure equations for incompressible fluid flow. In standard wave computations, a numerical wave flume is configured, where the computational domain is a numerical model of the laboratory experimental wave flume (see Fig. 3.3). The velocity and pressure in the domain must fulfill all the boundary conditions at the side and end walls, the bottom, as well as the free-surface or interface. Incident water waves are generated at the in-flow boundary end of the flume, and they are passed through the open boundary at the other end. In particular, the free-surface boundary condition is the most important condition to define when simulating wave motion.

A variety of computational methods to implement dynamic and kinematic free-surface conditions have been developed, which differ depending on the grid system, fundamental numerical scheme to solve the governing equations, and the surface detecting or tracking algorithm. Details of these techniques are explained in §4.2 for the volume-of-fluid (VOF) method and §5.4 for a general description of surface capturing schemes. Particle methods, described in Chapter 6, have unique definitions of fluid and solid boundaries when using gridless computations. Therefore, describing boundary conditions with gridless computations is quite different from other finite difference computations using discrete grids, which will also be dealt with in each chapter.

This section focuses on common procedures to impose boundary conditions for the finite difference wave computation.

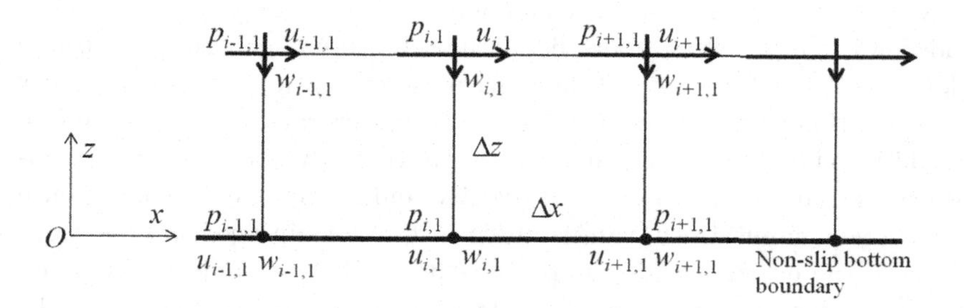

Fig. 3.4. Collocation grids near the boundary. All variables at the boundary grids are explicitly constrained by the wall boundary condition.

3.2.1. *Bottom and wall conditions*

The boundary condition for velocity $\boldsymbol{u} = (u, v, w)$ at an impermeable wall or bottom is given by

$$\boldsymbol{u} \cdot \boldsymbol{n} = 0 \tag{3.44}$$

where $\boldsymbol{n}$ is the unit normal vector. For instance, for the case with an impermeable bottom boundary located at $z = -h(x, y)$ in the coordinate system defined in Fig. 3.3 (the origin of the vertical axis z is located at the still water level), $\boldsymbol{n}$ is defined to be $(\partial h/\partial x, \partial h/\partial y, 1) / \sqrt{(\partial h/\partial x)^2 + (\partial h/\partial y)^2 + 1}$.

The tangential velocity also vanishes at the non-slip boundary

$$\boldsymbol{u} \cdot \boldsymbol{s} = 0 \tag{3.45}$$

where $\boldsymbol{s}$ is the unit tangential vector. In Cartesian coordinates with orthogonal horizontal axes x and y, and vertical axis z, Eqs. (3.44) and (3.45) for a flat bottom (constant depth) are reduced to

$$u = v = w = 0 \tag{3.46}$$

The computational procedure to impose the condition of Eq. (3.46) depends on the grid arrangement, whether it be a collocated or staggered arrangement. In the former, all variables are stored at the same grid points, so zero velocity can be explicitly given at grid locations that coincide with the boundary; this is a straightforward way to achieve the condition of Eq. (3.46) (see Fig. 3.4). A collocated grid system has many advantages in complicated domains, since the arbitrary shapes of the computing cells can be designed to fit the boundary in generalized coordinates.

A staggered arrangement is commonly used to compute incompressible fluid flows. The pressure nodes lie at the cell centers, and the cell-normal velocity is defined at each cell face, where a straightforward second-order approximation of the derivatives is attained by central finite difference (see Fig. 3.5). While this arrangement has significant advantages to reduce numerical oscillations of pressure and velocity, and to improve the convergence of iterative computations, values stored in the domain need to be extrapolated to the exterior grids to approximately fulfill the impermeable and nonslip conditions at the boundary. In the Cartesian staggered grids, the forward and backward Taylor approximations of the tangential (horizontal) velocity $u(z = \pm\Delta z/2)$ about $z = 0$ (i.e., boundary level) are

$$u(\Delta z/2) = u(0) + \frac{\Delta z}{2}\frac{\partial u(0)}{\partial z} + O\left(\frac{\Delta z^2}{4}\right) \tag{3.47}$$

$$u(-\Delta z/2) = u(0) - \frac{\Delta z}{2}\frac{\partial u(0)}{\partial z} + O\left(\frac{\Delta z^2}{4}\right) \tag{3.48}$$

The sum of the above equations provides the first-order approximation of the velocity extrapolated from the adjacent inner grid to the exterior of the domain

$$u(-\Delta z/2) = -u(\Delta z/2) \tag{3.49}$$

This condition indicates the horizontal velocity gradient $\partial u/\partial x|_+$ in the domain is opposite in sign to the extrapolated one $\partial u/\partial x|_- : \partial u/\partial x|_+ = -\partial u/\partial x|_-$. Thus the continuity condition in the cell adjacent to the boundary follows the relation

$$\begin{aligned}\frac{\partial w}{\partial z}|_+ &= -\frac{\partial u}{\partial x}|_+ \\ &= \frac{\partial u}{\partial x}|_- = -\frac{\partial w}{\partial z}|_-\end{aligned} \tag{3.50}$$

The same approximation provides the extrapolated w outside of the domain:

$$w(\Delta z) = w(0) + \Delta z\frac{\partial w(0)}{\partial z}|_+ + O\left(\frac{\Delta z^2}{4}\right) \tag{3.51}$$

$$w(-\Delta z) = w(0) - \Delta z\frac{\partial w(0)}{\partial z}|_- + O\left(\frac{\Delta z^2}{4}\right) \tag{3.52}$$

The normal (vertical) velocity $w(z = 0)$ can be simply given to be zero at the impermeable boundary. Substituting the relation of Eq. (3.50) into the above approximations, we have the boundary condition for w:

$$w(-\Delta z) = w(\Delta z) \tag{3.53}$$

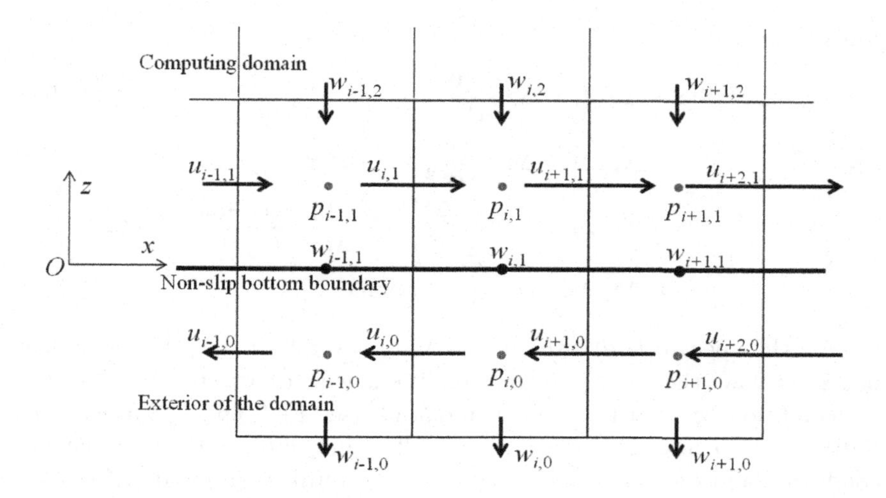

Fig. 3.5. Staggered grids near the non-slip boundary. Since there is no grid to define u and p at the boundary, unreal (ghost) variables for the exterior of the domain are given via extrapolation to satisfy the non-slip condition at the boundary.

Higher order approximations are possible if suitable velocity gradients (or shear) can be assumed at the boundary.

The boundary condition for pressure at the impermeable flat bottom is derived from the vertical momentum equation

$$\frac{\partial w}{\partial t}+u\frac{\partial w}{\partial x}+v\frac{\partial w}{\partial y}+w\frac{\partial w}{\partial z}=-\frac{1}{\rho}\frac{\partial p}{\partial z}+\nu\left(\frac{\partial^2 w}{\partial x^2}+\frac{\partial^2 w}{\partial y^2}+\frac{\partial^2 w}{\partial z^2}\right)-g \quad (3.54)$$

At the bottom boundary ($z=0$), the above equation can be reduced by the boundary condition of Eq. (3.46) to

$$\frac{\partial p(0)}{\partial z}=\mu\frac{\partial^2 w(0)}{\partial z^2}-\rho g \quad (3.55)$$

Again, the first order approximations of $p(\pm\Delta z/2)$ are used for the extrapolation:

$$p(\Delta z/2)=p(0)+\frac{\Delta z}{2}\frac{\partial p(0)}{\partial z}+O\left(\frac{\Delta z^2}{4}\right) \quad (3.56)$$

$$p(-\Delta z/2)=p(0)-\frac{\Delta z}{2}\frac{\partial p(0)}{\partial z}+O\left(\frac{\Delta z^2}{4}\right) \quad (3.57)$$

which leads to

$$\begin{aligned} p(-\Delta z/2) &\approx p(\Delta z/2) - \Delta z \frac{\partial p(0)}{\partial z} \\ &= p(\Delta z/2) - \Delta z \mu \frac{\partial^2 w(0)}{\partial z^2} + \rho g \Delta z \\ &\approx p(\Delta z/2) - \Delta z \mu \frac{w(\Delta z) - 2w(0) + w(-\Delta z)}{\Delta z^2} + \rho g \Delta z \\ &= p(\Delta z/2) - 2\mu \frac{w(\Delta z)}{\Delta z} + \rho g \Delta z \end{aligned} \tag{3.58}$$

For the special case without boundary layer flow, i.e. behavior like potential flow, the slip boundary condition, which corresponds to a mirror condition, is used for the extrapolation (see Fig. 3.6). Substituting $\partial u(0)/\partial z = 0$ into Eqs. (3.47) and (3.48), and $w(0) = 0$ (impermeable condition) and $\partial w(0)/\partial z|_+ = \partial w(0)/\partial z|_-$ into Eqs. (3.51) and (3.52),

$$u(-\Delta z/2) = u(\Delta z/2) \tag{3.59}$$

$$w(-\Delta z) = -w(\Delta z) \tag{3.60}$$

Thus the pressure for the slip condition is extrapolated by Eq. (3.58) with Eq. (3.60):

$$p(-\Delta z/2) = p(\Delta z/2) + \rho g \Delta z \tag{3.61}$$

The side wall conditions of a three-dimensional numerical wave flume can also be approximated by an identical approach.

For a non-slip vertical side wall, using the faithful model of a laboratory wave flume, we have

$$u(-\Delta y/2) = -u(\Delta y/2) \tag{3.62}$$

$$v(0) = 0 \tag{3.63}$$

$$v(-\Delta y) = v(\Delta y) \tag{3.64}$$

$$w(-\Delta y/2) = -w(\Delta y/2) \tag{3.65}$$

Since the ocean has no side walls, one may not want to compute the side wall boundary layer. For this case, the slip mirror condition may be used:

$$u(-\Delta y/2) = u(\Delta y/2) \tag{3.66}$$

$$v(0) = 0 \tag{3.67}$$

$$v(-\Delta y) = -v(\Delta y) \tag{3.68}$$

$$w(-\Delta y/2) = w(\Delta y/2) \tag{3.69}$$

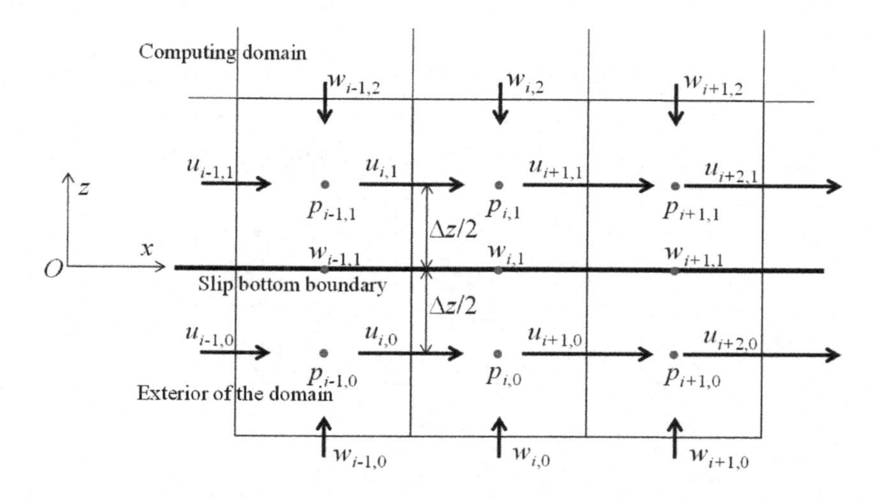

Fig. 3.6. Staggered grids near the slip boundary. The variables outside of the domain are also extrapolated to satisfy the slip condition at the boundary.

To simulate infinitely long-crested waves or laterally periodic flows, a periodic boundary condition is often adopted for both of the side walls of the numerical wave flume:

$$u(-\Delta y/2) = u(L - \Delta y/2) \tag{3.70}$$

$$v(0) = v(L) \tag{3.71}$$

$$v(-\Delta y) = v(L - \Delta y) \tag{3.72}$$

$$w(-\Delta y/2) = w(L - \Delta y/2) \tag{3.73}$$

where L is the width of the flume.

While the non-slip boundary should be implemented for the bottom and the structure wall in viscous flows, the grid width near the boundary needs to be carefully determined. For cases where the boundary layer is unresolved on coarse grids, the computed tangential viscous stress $\mu\partial u/\partial z$ numerically depends on the grid width adjacent to the boundary. Accordingly, the grid width becomes a parameter that may change the mechanical balance within the boundary layer to induce unexpectedly excessive momentum flux and form an unreasonably thick boundary layer. Since using highly resolved cells over the entire domain may be impractical and unrealistic in terms of computational cost, a suitable coordinate transform should be adopted to generate finer resolution for only the cells near the boundary.

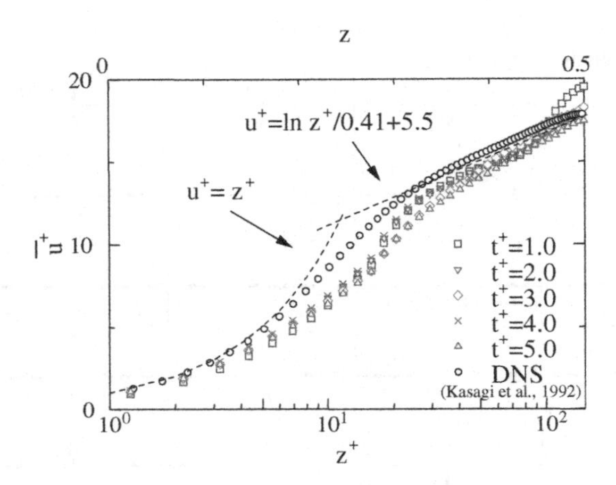

Fig. 3.7. An example of turbulent boundary layers computed by Large Eddy Simulations.[9]

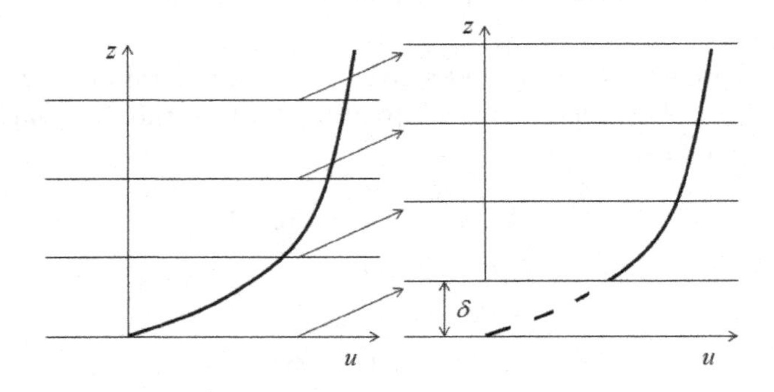

Fig. 3.8. Schematic illustration of the coordinate shift (TBL wall model).

Readers are referred to a variety of grid generation methods introduced in other specialized literature sources.[5]

In shallow water of the ocean, significant turbulence is produced near the seabed and coastal structures, resulting in a turbulent boundary layer (TBL). A TBL governs sediment transport on the coast and also affects wave energy dissipation. For these reasons, we need to properly account for TBLs in wave computations.

A TBL in a steady flow is known to generate multi-layer structures near the wall (see Fig. 3.7): a linear sublayer ($z^+ \leq 5$), a buffer layer ($5 < z^+ \leq$

40), a logarithmic region ($40 < z^+$ and $z < 0.2\delta$) and a velocity defect region ($40 < z^+$ to $z = \delta$), where the wall distance $z^+ = u_\tau z/\nu$, the friction velocity $u_\tau = \sqrt{\tau_w/\rho}$, the wall shear is τ_w, and the boundary layer thickness is δ. The computations to resolve every thin layer are very expensive, e.g., the viscous wall region involving the linear and buffer layers may be of $O(0.1$ mm) in thickness, and the outer layer may be slightly thicker but less than 10 mm on the laboratory flow scale. As already mentioned, low-resolution computations for these layers may cause unrealistic erroneous outputs. While turbulence computations for TBLs over sea surfaces on very finely resolved grids have been successfully performed[9] (see Fig. 3.7), some simple models to approximate the TBL also have been proposed to reduce computational expenses.

The simplest model is to slightly shift the grid system away from the boundary[10] (see Fig. 3.8). The first grid point can be placed in the logarithmic region, at which point the theoretical mean velocity ($u^+ = \ln z^+/\kappa + 5.5$, where κ is the Karman constant) and its derivatives can be given as the boundary condition. However, the classical TBL theory for steady flow is unable to describe several major properties of unsteady TBLs developed under waves: the temporal variation of the defect layer thickness, and the phase shift of the nearbed velocity and its gradient. Careful examination of the results are required when such a simple boundary model is used.

3.2.2. *In-flow conditions*

There are two types of methods to generate incident waves at the offshore boundary. A numerical wave generator oscillating an impermeable paddle with grid points can be incorporated in the numerical flume as a faithful model of a laboratory wave generator. Arbitrary incident waves can be designed using classical wave maker theory. This method was commonly used in a boundary element method that can flexibly configure moving boundary nodes (e.g., Xue *et al.*[6]). However, this method is out of favor for other finite difference, volume and element methods that use computing grids since the grids in the domain need to be transformed along with the moving paddle.

Instead, the theoretical velocity, surface elevation and pressure are generally implemented at the offshore boundary. With this method, suitable wave theories should be chosen to compute incident wave parameters such as water depth (h), wave height (H), wave period (T) and wavelength (L). In the real ocean, relatively small water waves are initially generated in deep water by wind; these may be approximated as linear Airy waves described by small amplitude wave theory (see §1.3.2 for the theoretical background of potential wave equations). They develop gradually from successive wind stresses and have a steeper form during propagation toward shore, which is well approximated as Stokes waves. In shallow water, the waves have significantly grown and become nonlinear in shape during shoaling, which is described by cnoidal wave theory.

Surface elevation, horizontal and vertical velocities, and pressure to be implemented at the boundary on the basis of major wave theories are given below (see Svensen[11] for details of the wave theories).

Airy waves

The simplest linear wave solutions are derived from small amplitude wave theory (see also §1.3.2).
⟨surface elevation⟩

$$\zeta_1 = \frac{H}{2}\cos(kx - \sigma t) \tag{3.74}$$

⟨horizontal velocity along a wave ray⟩

$$u_1 = \frac{H}{2}\sigma\frac{\cosh k(h+z)}{\sinh kh}\cos(kx - \sigma t) \tag{3.75}$$

⟨vertical velocity⟩

$$w_1 = \frac{H}{2}\sigma\frac{\sinh k(h+z)}{\sinh kh}\sin(kx - \sigma t) \tag{3.76}$$

⟨pressure⟩

$$\frac{p_1}{\rho g} = -z + \frac{H}{2}\frac{\cosh k(h+z)}{\cosh kh}\cos(kx - \sigma t) \tag{3.77}$$

Here, the wave number $k = 2\pi/L$ and angular frequency $\sigma = 2\pi/T$, which follow a dispersion relation $\sigma^2 = gk\tanh kh$.

Second-order Stokes waves

Additional nonlinear wave components to the linear components (ζ_1, u_1, w_1, p_1) express a steeper wave form.

⟨surface elevation⟩

$$\zeta_2 = \zeta_1 + \frac{H^2 k}{16}\frac{\cosh kh}{\sinh^3 kh}(2 + \cosh 2kh)\cos 2(kx - \sigma t) \tag{3.78}$$

⟨horizontal velocity⟩

$$u_2 = u_1 + \frac{3}{16}H^2\sigma k\frac{\cosh 2k(h+z)}{\sinh^4 kh}\cos 2(kx-\sigma t) \tag{3.79}$$

⟨vertical velocity⟩

$$w_2 = w_1 + \frac{3}{16}H^2\sigma k\frac{\sinh 2k(h+z)}{\sinh^4 kh}\sin 2(kx-\sigma t) \tag{3.80}$$

⟨pressure⟩

$$\frac{p_2}{\rho g} = p_1 + \frac{kH^2}{8\sinh 2kh}\left\{\left(3\frac{\cosh 2k(h+z)}{\sinh^2 kh}-1\right)\cos 2(kx-\sigma t) + 1 - \cosh 2k(h+z)\right\} \tag{3.81}$$

Cnoidal waves

Fluid motion in steep waves in shallow water is described by the cnoidal wave solutions in terms of Jacobi elliptic functions, sn, cn and dn.

⟨surface elevation⟩

$$\frac{\zeta_c}{h} = \frac{a}{h}\left(1-\frac{3}{4}\frac{a}{h}\right)\mathrm{cn}^2(\alpha x) + \frac{3}{4}\left(\frac{a}{h}\right)^2\mathrm{cn}^4(\alpha x) \tag{3.82}$$

⟨horizontal velocity⟩

$$\begin{aligned}\frac{u_c}{\sqrt{gh}} &= 1 + \frac{2m-1}{2m}\frac{a}{h} - \frac{21m^2-6m-9}{40m^2}\left(\frac{a}{h}\right)^2 \\ &+ \frac{3(1-m)}{4m}\left(\frac{2z}{h}+\frac{z^2}{h^2}\right)\left(\frac{a}{h}\right)^2 \\ &- \left\{1-\frac{a}{h}\frac{7m-2}{4m}-\frac{a}{h}\frac{6m-3}{2m}\left(\frac{2z}{h}+\frac{z^2}{h^2}\right)\right\}\left(\frac{a}{h}\right)\mathrm{cn}^2(\alpha x) \\ &- \left\{\frac{5}{4}+\frac{9}{4}\left(\frac{2z}{h}+\frac{z^2}{h^2}\right)\right\}\left(\frac{a}{h}\right)^2\mathrm{cn}^4(\alpha x)\end{aligned} \tag{3.83}$$

⟨vertical velocity⟩

$$\frac{w_c}{\sqrt{gh}} = -\sqrt{\frac{3}{m}\left(\frac{a}{h}\right)^3}\left(1+\frac{z}{h}\right)\mathrm{cn}(\alpha x)\mathrm{sn}(\alpha x)\mathrm{dn}(\alpha x) \tag{3.84}$$

$$\left\{1-\frac{5m+2}{8m}\frac{a}{h}-\frac{2m-1}{2m}\left(\frac{2z}{h}+\frac{z^2}{h^2}\right)\frac{a}{h}-\left(\frac{1}{2}-\frac{3z}{h}-\frac{3}{2}\frac{z^2}{h^2}\right)\mathrm{cn}^2(\alpha x)\right\}$$

⟨pressure⟩

$$\frac{p_c}{\rho g h} = \frac{\zeta_c - z}{h} + \frac{3}{4m}\left(\frac{2z}{h} + \frac{z^2}{h^2}\right)\left(\frac{a}{h}\right)^2 \tag{3.85}$$
$$\{-1 + m - 2(2m-1)\mathrm{cn}^2(\alpha x) + 3m\mathrm{cn}^4(\alpha x)\}$$

Here, $\alpha = \frac{1}{h}\sqrt{\frac{3}{4m}\left(\frac{a}{h}\right)}\left\{1 + \frac{7m-2}{8m}\left(\frac{a}{h}\right)\right\}^{-1}$, written in terms of the Jacobi elliptic function parameter m, and a is the amplitude ($= H/2$). It should be noted that a cnoidal wave asymptotically approaches a solitary wave when $m \to 1$ and an Airy wave when $m \to 0$.

Practical applications of these boundary conditions are also explained in §4.3.3.

3.2.3. *Open boundary*

To avoid forming a multiple reflection system, the onshore boundary should be set as an open boundary in the numerical wave flume. An exception would be surf-zone simulations, where wave energy is dissipated by wave breaking in the domain, and no waves are transmitted back toward the onshore boundary. However, appropriate boundary conditions that can perfectly transmit arbitrary wave fields, namely irregular waves and nonlinear waves, may not yet be known.

As the simplest out-flow condition, the zero normal velocity gradient condition has been used in uni-directional flow (i.e., not the wave field)

$$\frac{\partial \boldsymbol{u}}{\partial n} = 0 \tag{3.86}$$

This condition may work to transmit a part of the momentum and somewhat reduce the unwanted reflection that occurs when waves propagate in a direction perpendicular to the boundary face. However, when incident waves propagate toward the boundary at an oblique angle, the zero normal velocity gradient condition may give poor results. And, this condition may give limited effects for unsteady flows in a wave field.

Orlanski[7] proposed a general radiation condition based on the well-known Sommerfeld's condition to define the variable φ at the boundary, which follows

$$\frac{\partial \varphi}{\partial t} + c\frac{\partial \varphi}{\partial n} = 0 \tag{3.87}$$

where c is the phase velocity approximately determined by the finite difference at the previous time step

$$c = -\frac{\delta\varphi}{\delta t} / \frac{\delta\varphi}{\delta n} \tag{3.88}$$

Here, δ is the finite difference operator. To avoid a singular computation, the following restriction of c is used.

$$c = \begin{cases} 0 & (c < 0) \\ c & (0 \le c \le \Delta x/\Delta t) \\ \Delta x/\Delta t & (c > \Delta x/\Delta t) \end{cases} \tag{3.89}$$

The computed c is substituted into Eq. (3.87) to update φ in the upwind finite difference.

A multi-dimensional version of Orlanski's model has been considered by Raymond and Kuo.[8] In three-dimensional space, the condition in Eq. (3.87) takes the form

$$\frac{\partial\varphi}{\partial t} = -\mathbf{c} \cdot \nabla\varphi = -\left(c_x \frac{\partial\varphi}{\partial x} + c_y \frac{\partial\varphi}{\partial y} + c_z \frac{\partial\varphi}{\partial z}\right) \tag{3.90}$$

where $\mathbf{c}$ is the phase velocity vector, and c_x, c_y, c_z are the projections of $\mathbf{c}$ in the x, y, z directions. The finite difference form of $\mathbf{c}$ becomes

$$\mathbf{c} = -\frac{\delta\varphi}{\delta t}\left(\frac{\delta\varphi}{\delta x}, \frac{\delta\varphi}{\delta y}, \frac{\delta\varphi}{\delta z}\right)\left\{\left(\frac{\delta\varphi}{\delta x}\right)^2 + \left(\frac{\delta\varphi}{\delta y}\right)^2 + \left(\frac{\delta\varphi}{\delta z}\right)^2\right\}^{-1} \tag{3.91}$$

which is also calculated using φ at the previous time step. The identical restriction of Eq. (3.89) is adopted for each component of $\mathbf{c}$. Using the computed $\mathbf{c}$, new values of φ at the next time step can be estimated from Eq. (3.90) via the finite difference method.

An absorbing sponge zone (layer) is commonly used to attenuate waves before they arrive at the shoreward boundary. In this method, additional unphysical viscosity or drag, which is monotonically increased in the absorbing zone adjacent to the shoreward boundary, causes gradual attenuation of wave energy. This method requires an additional domain for the sponge zone, where the length should be comparable to the transmitted wavelength. A combination of this type of absorbing model with another open boundary condition significantly reduces the unwanted reflection (see also §4.3.4 for the practical applications of these boundary conditions).

3.3. Solution and Procedures for the Pressure Equation

A set of governing equations, the Navier–Stokes equation and the Poisson pressure equation, is integrated to model the evolution of fluid behavior. The Poisson equation does not have a partial difference for time, therefore, an iterative solver is necessary to satisfy the incompressibility of the fluid. This section describes numerical methods to solve the matrix of the Poisson equation. An overview of several matrix solvers will be illustrated, and a detailed mathematical description is referenced in the bibliography at the end of this chapter.

Besides an advection equation, a solution for elliptic differential equations such as the Poisson equation and the Laplace equation is determined with the lateral boundary condition. The physical meaning of this condition relates to neglecting the acoustic velocity. An infinite acoustic velocity gives an infinite propagation speed for pressure, and this kind of field can be described with the Poisson equation.

The Poisson equation in a two dimensional domain is

$$\frac{\partial^2 p}{\partial x^2} + \frac{\partial^2 p}{\partial y^2} = f(x, y) \tag{3.92}$$

where p is pressure and f is an arbitrary scalar function. The finite difference approximation of the Poisson equation by central difference scheme can be derived as

$$\frac{p_{i-1,j} - p_{i,j} + p_{i+1,j}}{(\Delta x)^2} + \frac{p_{i,j-1} - p_{i,j} + p_{i,j+1}}{(\Delta y)^2} = f_{i,j} \tag{3.93}$$

where Δx and Δy are the mesh widths in the x and y directions, i and j are the discrete location numbers in the x and y directions that correspond to discrete locations x_i and y_j, respectively. A set of linear relations by Eq. (3.93) can be derived with a number of discrete points. Solving simultaneous linear equations with a lateral boundary condition of p and internal values of $p_{i,j}$ can be computed as

$$\boldsymbol{A}\boldsymbol{x} = \boldsymbol{b} \tag{3.94}$$

where $\boldsymbol{A}$ is the coefficient matrix from Eq. (3.93), $x_k = p_{i,j} = (p_{11}, p_{21}, \cdots, p_{NN})^T$, and $\boldsymbol{b}$ is the lateral boundary condition. The numerical solution of an elliptic differential equation is basically independent of the mesh resolution. Therefore, to maintain low-cpu costs for the numerical solver of Eq. (3.93), computing the inverse matrix $\boldsymbol{A}$ becomes important.

As indicated by Eq. (3.93) and Eq. (3.94), the coefficient matrix $\boldsymbol{A}$ is determined by the numerical scheme of the finite difference scheme. For the two dimensional Poisson equation, the information that $\boldsymbol{A}$ at i, j gives for locations $i \pm n, j \pm n$ depends on the scheme. Therefore, the coefficient matrix $\boldsymbol{A}$ becomes a band matrix in the diagonal direction as shown in Fig. 3.9(a). As a result, the coefficient matrix $\boldsymbol{A}$ can be regarded as a sparse matrix that contains zero components for most of the matrix except the diagonal components, as shown in Fig. 3.9(b). This characteristic of being a sparse matrix generally applies whether the governing equation expands to three dimensions, or whether a higher-order finite difference scheme is used. The sparse band matrix can be solved quickly if the appropriate numerical solver is used. Most of the computational cost to solve the incompressible Navier–Stokes equation depends on solving the Poisson equation; therefore, choosing the appropriate numerical scheme for the Poisson equation is important to reduce the total cost. On the other hand, aside from computational cost, the numerical solution is the same even if a different numerical scheme is used for the Poisson equation. Thus, using a general numerical package such as NAG[a], IMSL[b] or LANPACK[c] is reasonable for a beginner, although the particular solver one chooses is important, as described in the following section.

3.3.1. *Matrix solvers*

Numerical methods to solve simultaneous linear equations can be classified into three categories: a direct method, an iteration method, and a conjugate gradient method. The direct method obtains an exact solution, but the computational cost for an $N \times N$ size matrix is expensive, since the cost is proportional to N^3. The iteration method starts with an initial value and repeats iterations until it converges on a solution. The iteration method was widely used at the beginning of computational fluid mechanics. The conjugate gradient method falls between the two methods, and it has an advantage in the computational cost over the other two methods. Three different matrix solving methods will be illustrated in the following section.

3.3.1.1. *Direct method*

The algorithm for the direct method is robust and obtains a solution with certainty. However, the computational cost is proportional to the power to N^3 for an $N \times N$ size matrix. This method requires longer computational

[a]NAG, Numerical Algorithms Group
[b]IMSL, Visual Numerics Inc.
[c]Linear Algebra PACKageCopen library by netlib

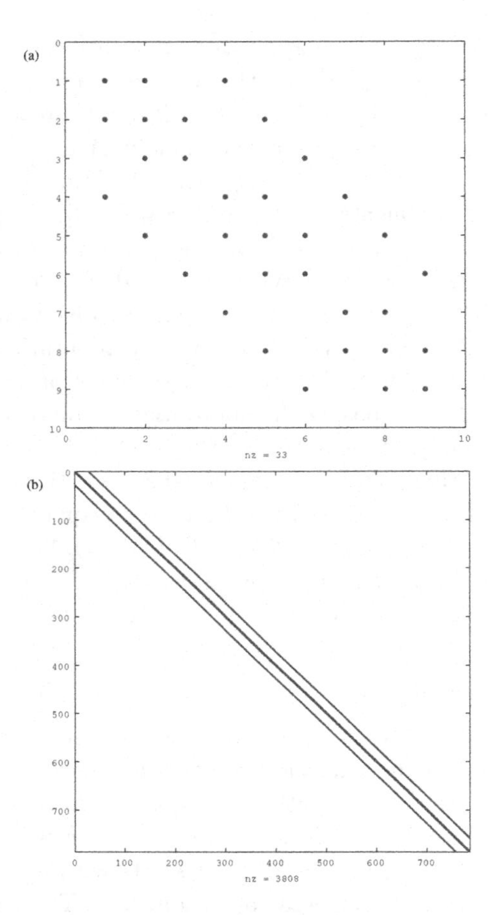

Fig. 3.9. Matrix of two dimensional Poisson equation; magnified view (top), over view (bottom).

times and is not appropriate for larger domain sizes. A representative example of the direct method is Gaussian elimination, solving the matrix as follows. For a given matrix $\boldsymbol{A}$,

$$\boldsymbol{A} = \begin{pmatrix} A_{11} & A_{12} & A_{13} & \cdots & A_{1N} \\ A_{21} & A_{22} & A_{23} & \cdots & A_{2N} \\ \vdots & \vdots & \vdots & \vdots & \vdots \\ A_{N1} & A_{N2} & A_{N3} & \cdots & A_{NN} \end{pmatrix} \tag{3.95}$$

multiply the results of A_{21}/A_{11} by the fist row and subtract the second row. Iterate this procedure from the first column to column $N-1$, and the matrix $\boldsymbol{A}$ can be rewritten as an upper triangular matrix $\boldsymbol{U}$

$$\boldsymbol{U} = \begin{pmatrix} A_{11} & A_{12} & A_{13} & \cdots & A_{1N} \\ 0 & A_{22} & A_{23} & \cdots & A_{2N} \\ \vdots & \vdots & \vdots & \vdots & \vdots \\ 0 & 0 & 0 & \cdots & A_{NN} \end{pmatrix} \tag{3.96}$$

This procedure is called a forward elimination. The obtained upper triangular matrix $\boldsymbol{U}$ can be solved starting from the bottom and proceeding to the top, and the solution of Eq. (3.94) is obtained at the end.

$$x_i = \frac{b_i - \sum_{j=i+1}^{N} A_{ij} x_j}{A_{ii}}, i \neq N \tag{3.97}$$

$$x_N = \frac{b_N}{A_{NN}} \tag{3.98}$$

The Gauss elimination method is quite general, and it can solve any type of simultaneous linear equations. However, the coefficient matrix for the finite difference scheme of the Poisson equation is a sparse matrix which generally has only diagonal components. Therefore, the Gauss elimination method is too expensive to solve this type of matrix. In numerical modeling of fluids, an appropriate method to solve the sparse matrix is the conjugate gradient method and the relaxation method.

3.3.1.2. *Iterative method*

The iterative method obtains a numerical solution by the iterative convergence of a simple algorithm; this method has been widely used since the 1970s. The linear iteration method gives the following relationship between the solution $\boldsymbol{x}$ at a present step n and a previous step $n-1$

$$\boldsymbol{x}^n = \boldsymbol{x}^{n-1} + \boldsymbol{R}(\boldsymbol{b} - \boldsymbol{A}\boldsymbol{x}^{n-1}), n \geq 1 \tag{3.99}$$

where $\boldsymbol{R}$ is an acceleration coefficient matrix. As the right hand side of $\boldsymbol{x}^{n-1}$ converges toward the solution, $\boldsymbol{b} - \boldsymbol{A}\boldsymbol{x}^{n-1}$ approaches zero and the appropriate solution can be obtained with an error of $\varepsilon = \| \boldsymbol{b} - \boldsymbol{A}\boldsymbol{x}^{n-1} \|$. The choice of acceleration coefficient is important for this method. There are several types of iteration methods, and some are briefly introduced here.

Gauss–Seidel method. The Gauss–Seidel method decomposes the matrix $\boldsymbol{A}$ to estimate the acceleration coefficient $\boldsymbol{R}$.

$$\boldsymbol{A} = \boldsymbol{U} + \boldsymbol{D} + \boldsymbol{L} \tag{3.100}$$

$$\boldsymbol{R} = (\boldsymbol{L} + \boldsymbol{D})^{-1} \tag{3.101}$$

where $\boldsymbol{U}$, $\boldsymbol{D}$, and $\boldsymbol{L}$ are upper triangular components, diagonal components, and lower triangular components, respectively. Substituting Eq. (3.101) into Eq. (3.99) gives

$$\boldsymbol{x}^n = \boldsymbol{x}^{n-1} + \boldsymbol{D}^{-1}(\boldsymbol{b} - \boldsymbol{L}\boldsymbol{x}^n - \boldsymbol{D}\boldsymbol{x}^{n-1} - \boldsymbol{U}\boldsymbol{x}^{n-1}) \tag{3.102}$$

Eq. (3.102) is iterated until the solution converges sufficiently as $\boldsymbol{\varepsilon} = \boldsymbol{x}^n - \boldsymbol{x}^{n-1}$, where $\boldsymbol{\varepsilon}$ is the magnitude of convergence. The Gauss–Seidel method is the most basic method in the iterative method.

Successive Over-Relaxation (SOR) method. A faster modification for the Gauss–Seidel method was proposed: the Successive Over-Relaxation method (SOR method). This is the first choice of an iterative method for solving matrices. The SOR method introduces an over-relaxation factor ω for Eq. (3.101).

$$\boldsymbol{R} = (\boldsymbol{L} + \omega^{-1}\boldsymbol{D})^{-1} \tag{3.103}$$

Substituting Eq. (3.103) into (3.99) gives

$$\boldsymbol{x}^n = (\boldsymbol{D} + \omega\boldsymbol{L})^{-1}\{\omega\boldsymbol{b} + [(1-\omega)\boldsymbol{D} - \omega\boldsymbol{U})\boldsymbol{x}^{n-1}\} \tag{3.104}$$

$$= \boldsymbol{x}^{n-1} + \omega\boldsymbol{D}^{-1}(\boldsymbol{b} - \boldsymbol{L}\boldsymbol{x}^n - \boldsymbol{D}\boldsymbol{x}^{n-1} - \boldsymbol{U}\boldsymbol{x}^{n-1}) \tag{3.105}$$

Note that when ω equals zero, the SOR method is equivalent to the Gauss–Seidel method. The optimum acceleration coefficient ω is proposed theoretically. For example, a square matrix with size $N \times N$ has an optimum ω value $\omega_{opt} = 2/(1 + \sin(\pi/N))$. The algorithm of the SOR method is simple to implement and easy to code. Therefore, it has been widely used for computational fluid mechanics.

Alternating Direction Implicit (ADI) method. To decompose matrix $\boldsymbol{A}$, the two-dimensional form of partial differential equation (3.100) can be divided into x and y (or i and j in discretized coordinates) directions as

$$\boldsymbol{A} = \boldsymbol{U}_1 + \boldsymbol{D}_1 + \boldsymbol{L}_1 + \boldsymbol{U}_2 + \boldsymbol{D}_2 + \boldsymbol{L}_2 \tag{3.106}$$

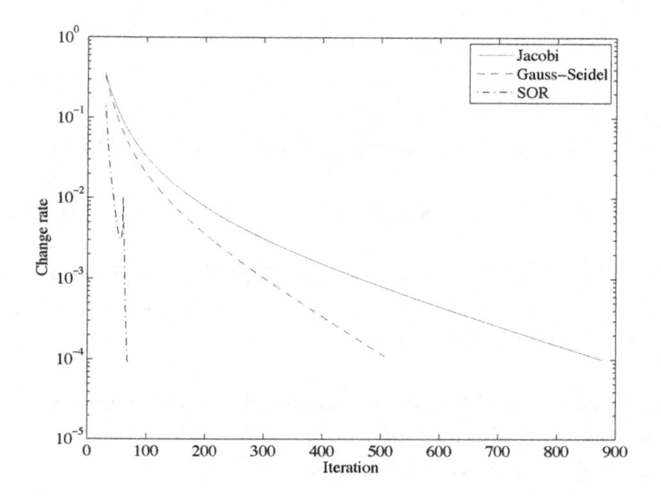

Fig. 3.10. Examples of convergence on a solution.

where subscripts 1 and 2 denote the x and y components, respectively. The ADI method uses the relation of Eq. (3.106) as

$$\boldsymbol{x}^{n-1/2} = \boldsymbol{x}^{n-1} + (\boldsymbol{U}_1 + \boldsymbol{D}_1 + \boldsymbol{L}_1 + \omega \boldsymbol{I})^{-1}(\boldsymbol{b} - \boldsymbol{A}\boldsymbol{x}^{n-1}) \quad (3.107)$$

$$\boldsymbol{x}^{n} = \boldsymbol{x}^{n-1/2} + (\boldsymbol{U}_2 + \boldsymbol{D}_2 + \boldsymbol{L}_2 + \omega \boldsymbol{I})^{-1}(\boldsymbol{b} - \boldsymbol{A}\boldsymbol{x}^{n-1/2}) \quad (3.108)$$

where $\boldsymbol{I}$ is the unit matrix. The ADI method is classified as a multi-step method. Due to the assumption of freezing in one direction, each discrete step should be small. The computational cost is proportional to $\sqrt{N}$, which is less than the direct method and suitable for parallel computation.

3.3.1.3. *Conjugate Gradient method*

The Conjugate Gradient method (CG method) is a relatively new method that was originally developed in the 1960s. In the 1980s, a precondition was introduced to the CG method, and it became widely used to solve large-sized sparse matrices. The CG method is one of the iterative methods that guarantees convergence within a finite number of iterations. The computational cost is linearly proportional to the size of matrix, or length N, and therefore it costs less than the other methods mentioned here.

The basic idea of the CG method starts by assuming an arbitrary function $f(\boldsymbol{x})$ for $\boldsymbol{x}$ as

$$f(\boldsymbol{x}) = \frac{1}{2}\boldsymbol{x} \cdot \boldsymbol{A}\boldsymbol{x} - \boldsymbol{x} \cdot \boldsymbol{b} \tag{3.109}$$

Finding the minimum limit of $f(\boldsymbol{x})$ is equivalent to obtaining the solution of $\boldsymbol{x}$ for the matrix $\boldsymbol{A}$. Thus differentiation of Eq. (3.109) gives

$$\boldsymbol{\nabla} f(\boldsymbol{x}) = \boldsymbol{A}\boldsymbol{x} - \boldsymbol{b} \tag{3.110}$$

and the solution $\boldsymbol{x}'$ for the condition of $\boldsymbol{\nabla} f(\boldsymbol{x}') = 0$ gives the optimum solution of $\boldsymbol{A}\boldsymbol{x}' = \boldsymbol{b}$. The numerical procedure of the CG method can be summarized as follows.

(1) Determine the initial value $\boldsymbol{x}^0$ and search direction $\boldsymbol{p}^0$
(2) Calculate α to minimize $f(\boldsymbol{x}^0 + \alpha^0\boldsymbol{p}^0)$
(3) Regard an approximated solution as $\boldsymbol{x}^1 = \boldsymbol{x}^0 + \alpha^0\boldsymbol{p}^0$
(4) Iterate steps (1)–(3) until the solution converges ($\| \boldsymbol{x}^1 - \boldsymbol{x}^0 \| \leq \varepsilon$)

The key point is to determine the searching direction $\boldsymbol{p}$. For the CG method, the searching direction $\boldsymbol{p}$ is determined with the following set of equations.

$$\alpha^{n-1} = \frac{\boldsymbol{r}^{n-1} \cdot \boldsymbol{r}^{n-1}}{\boldsymbol{p}^{n-1} \cdot \boldsymbol{A}\boldsymbol{p}^{n-1}} \tag{3.111}$$

$$\boldsymbol{x}^n = \boldsymbol{x}^{n-1} + \alpha^{n-1}\boldsymbol{p}^{n-1} \tag{3.112}$$

$$\boldsymbol{r}^n = \boldsymbol{r}^{n-1} - \alpha^{n-1}\boldsymbol{A}\boldsymbol{x}^{n-1} \tag{3.113}$$

where $\boldsymbol{r}^{n-1}$ is the residual vector of the solution $\boldsymbol{r} = \boldsymbol{b} - \boldsymbol{A}\boldsymbol{x}$ at iteration step $n-1$. If $\boldsymbol{r}^n$ is not enough small, continue following procedure.

$$\beta^{n-1} = -\frac{\boldsymbol{r}^n \cdot \boldsymbol{r}^n}{\boldsymbol{r}^{n-1} \cdot \boldsymbol{r}^{n-1}} \tag{3.114}$$

$$\boldsymbol{p}^n = \boldsymbol{r}^n + \beta^{n-1}\boldsymbol{p}^{n-1} \tag{3.115}$$

Note that $\boldsymbol{p}^n$ is orthogonal to $\boldsymbol{p}^n \cdot \boldsymbol{A}\boldsymbol{p}^{n-1}$. This formulation guarantees convergence of $\boldsymbol{r}^n = 0$ for $\boldsymbol{x}^n$ within n iterations.

A Preconditioned Conjugate Gradient method (PCG method) is the CG method with a preconditioned matrix. Preconditioning matrix $\boldsymbol{A}$ gives faster convergence on the solution, and it is an efficient method to obtain numerical solutions for large matrices.

BiCG method. A modified CG method for an asymmetric matrix $\boldsymbol{A}$ is the Bi-Conjugate Gradient method (BiCG method; Fletcher, 1976). The BiCG method is one of the gradient squared methods, and this method is characterized by the generation of search matrices $\boldsymbol{p}$ and $\boldsymbol{r}$. The BiCG method solves matrix $\boldsymbol{A}$ with its transpose, $\boldsymbol{A}^T$. The BiCG method is numerically unstable; therefore, the Bi-Conjugate Gradient Stabilized method (BiCGStab method) was developed, which is more numerically stable.

ICCG method. For asymmetric matrices, alternative PCG methods are the Incomplete Cholesky Conjugate Gradient method (ICCG method) and the Generalized Minimal RESidual method (GMRES method). The ICCG method uses the relation

$$\boldsymbol{A} = \boldsymbol{L}\boldsymbol{L}^T + \boldsymbol{E} \tag{3.116}$$

to keep the characteristics of a sparse matrix where $\boldsymbol{L}$ is the lower triangular matrix, $\boldsymbol{L}^T$ is its transpose, and $\boldsymbol{E}$ is the residual matrix. When using the residual matrix without the complete Cholesky decomposition, this type of method is called an incomplete Cholesky precondition. The simultaneous linear equations with the incomplete Cholesky precondition can be divided into upper and lower triangular matrices. The ICCG method has lower computational costs, and the computations become faster if diagonal components are weighted for accelerated convergence (Modified ICCG method, MICCG method).

3.3.1.4. *Summary of matrix solvers*

The numerical methods to solve simultaneous linear equations can be classified into three categories: direct method, iteration method, and conjugate gradient method. The different algorithms have advantages and disadvantages over the others. In addition, there are two choices for coding the matrix solver: either write it yourself or use a solver from a numerical library. A relatively easier algorithm to code is the SOR method and its computational cost is reasonable. The preconditioned conjugate gradient method is the best choice for large matrices (scale) with huge computations, coupled with a solver from a numerical library. However, spatial dimensions and finite difference schemes give different sparse matrices, and these conditions affect the number of iterations and convergence to a solution.

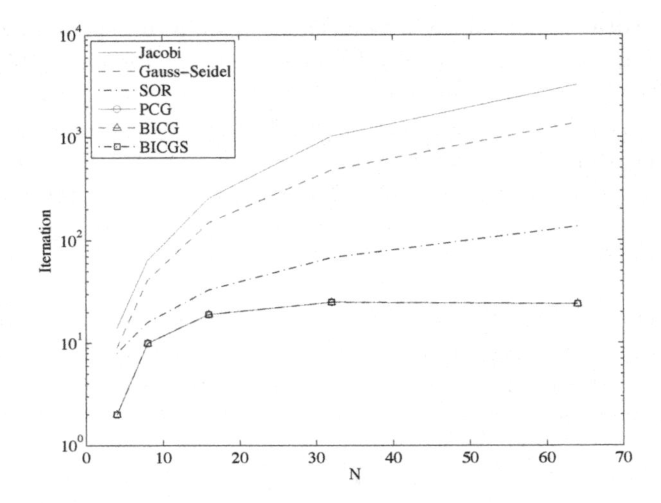

Fig. 3.11. Relationship between matrix size N and number of iterations (solid line: Jacobi, dashed line:Gauss–Seidel, dotted-dashed line: SOR method, o:PCG method, $\triangle$: BiCG method, $\square$: BiCGstab method).

3.3.2. *Example of solving matrix*

A series of numerical tests is examined for the two dimensional Poisson equation Eq. (3.93). The convergences to a solution are compared with the Jacobi method, the Gauss–Seidel method, the SOR method, the preconditional conjugate gradient method (PCG), and the BiCG method. The matrix is a square matrix, and the lateral boundary conditions are the Dirichlet boundary condition, with the value one for one edge, and zero for the rest of the edges.

Figure 3.10 shows the number of iterations to convergence using the different methods for an $N = 64$ square matrix. The convergences with the Jacobi method and the Gauss–Seidel method are relatively slower than for the SOR method; the computational costs for the two former methods are high. The convergence on a numerical solution with the SOR method is not monotonic, and it changes during the iteration process. These characteristics of the SOR method are also remarkable for cases with matrices of finer mesh.

Figure 3.11 shows the relationship between the number of iterations and matrix size for the case of a fixed residual as $\varepsilon = 10^{-6}$. The direct method requires more iterations, three different gradient conjugate methods are two orders of magnitude faster, and the SOR method is in between these methods. As shown in Fig. 3.11, the conjugate gradient method is appropriate for large matrices and in parallel computations. For these reasons, the conjugate gradient method is widely used for computational fluid dynamics problems.

References

1. Ferziger J.H. and Peric M. (2001): *Computational Methods for Fluid Dynamics*, 3rd Edition, Springer.
2. Garcia A. (1999): *Numerical Methods for Physics*, 2nd Edition Benjamin Cummings.
3. Fletcher C.A.J. (1991): *Computational Techniques for Fluid Dynamics: Fundamental and General Techniques*, Volume 1, Springer-Verlag.
4. Fletcher R. (1976): Conjugate Gradient Methods for Indefinite Systems, in Numerical Analysis Dundee 1975, Eds. by Watson,G., *Lecture Notes in Mathematics*, 506, Springer-Verlag, pp.73-89.
5. Thompson J.F., Soni B.K. and Weatherill N.P. (1998): *Handbook of grid generation*, CRC press.
6. Xue M., Xu H., Liu Y. and Yue D.K.P. (2001): Computations of fully nonlinear three-dimensional wave-wave and wave-body interactions. Part 1. Dynamics of steep three-dimensional waves, *J. Fluid Mech.*, Vol.438, pp.11-39.
7. Orlanski I. (1976): A simple boundary condition for unbounded hyperbolic flows, *J. Comp. Phys.*, Vol.21, pp.251-269.
8. Raymond W.H. and Kuo H.L. (1984): Aradiation boundary condition for multi-dimensional flows, *Quart. J. R. Met. Soc.*, Vol.110, pp.535-551.
9. Mutsuda H. (2001): Three-dimensional computation of wind-wave turbulent boundary layer flow with wave breaking, *Proc. Coastal Eng.*, JSCE, Vol.48, pp.61-65 (in Japanese).
10. Sagaut P. (2000): *Large eddy simulation for incompressible flows*, Springer.
11. Svensen Ib A. (2006): I*ntroduction to neashore hydrodynamics*, World Scientific.

Chapter 4

VOF Method

Koji Kawasaki

One of the main difficulties encountered with numerical simulations of free surface flows is how to treat the free surface configuration. Some interface capturing methods in the Eulerian representation, such as the height function method, line segment method, marker particle method, and volume of fluid (VOF) method, have been developed to represent the free surface appropriately. In this chapter, these interface capturing methods are briefly explained in §4.1, and the details of the VOF method are described in §4.2. A numerical wave flume (NWF), CADMAS-SURF, based on the VOF method is introduced in §4.3 with some practical examples of its applications in §4.4.

4.1. Interface Capturing Method

Some conventional interface/free-surface capturing methods are outlined below. General interpretations for the free-surface computation are also presented in §5.4.

4.1.1. *Height function method*

The height function method[13,18] is the simplest technique to represent and configure a free boundary. The free surface is defined by introducing a height function $H(x, y, t)$ at a distance from a reference line, as shown in Fig. 4.1(a). The dynamic behavior of the free surface can be estimated by calculating the time evolution equation of the height function (Eq. (4.1)), expressing the free surface kinematic boundary condition. However, this method does not work sufficiently when the local slope of the free surface exceeds the aspect ratio of the mesh cell, or when the free surface becomes

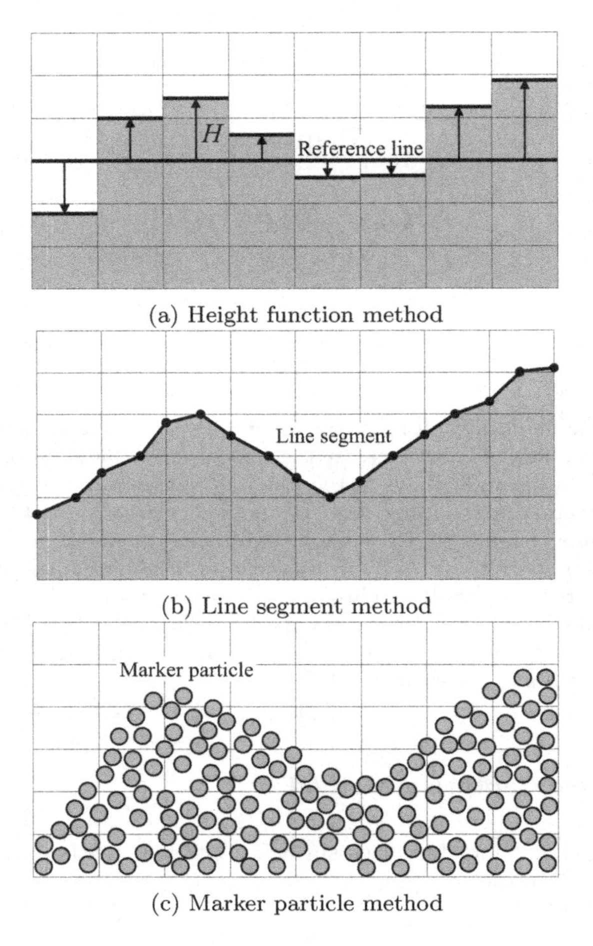

(a) Height function method

(b) Line segment method

(c) Marker particle method

Fig. 4.1. Interface capturing methods in Eulerian representation.

a multiple-valued surface, both of which occur with a plunging breaker.

$$\frac{\partial H}{\partial t} + u\frac{\partial H}{\partial x} + v\frac{\partial H}{\partial y} = w \tag{4.1}$$

where u, v, and w are fluid velocity components in the x, y and z coordinates, and t is time.

4.1.2. *Line segment method*

The line segment method is a generalization of the height function method with chains of short line segments or points connected by line segments,[17] as depicted in Fig. 4.1(b). The lengths of the segments have to be smaller than the minimum mesh size. Furthermore, this method requires slightly more computational cost as compared with the height function method. The time evolution of a chain of line segments is simply calculated by moving each point using the local fluid velocity. However, this method has a serious drawback when two surfaces intersect or when a surface folds over on itself. Also, extending the line segment method to a three-dimensional field is not easy.[19]

4.1.3. *Maker particle method*

In a marker particle method, the free surface dynamics are evaluated by tracking each particle in a fluid region that moves with the local fluid velocity,[11,21] as indicated in Fig. 4.1(c). Therefore, this method eliminates problems associated with intersecting surfaces, which are recognized problems with the height function and line segment methods. But, large amounts of computational costs are required due to the vast number of marker particles used in numerical simulations. Free surfaces are defined as a mesh cell containing markers, which connects with at least one neighboring cell with no markers. The actual location of the free surface in each cell has to be determined by an additional computation based on the distribution of markers within the cell. This method can be extended to a three-dimensional field, provided that sufficient storage requirements are available.[10]

4.1.4. *VOF method*

Nichols and Hirt[19] and Hirt and Nichols[12] proposed a VOF method to reduce memory storage requirements. The VOF method provides free surface information similar to the marker particle method by introducing only one additional function for each cell. This detail is described in the next section.

4.2. VOF Method

4.2.1. *Concept of the VOF method*

The VOF method relies on a Lagrangian description to track the volume of fluid. The equation to track a certain physical quantity $F(\boldsymbol{x}, t)$ in the Lagrangian description is generally expressed as

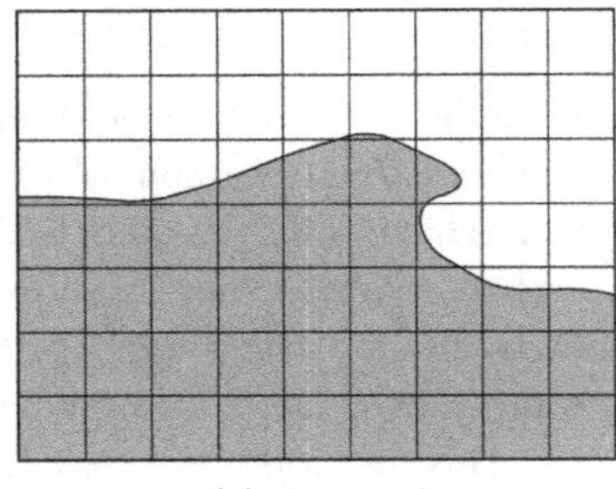

(a) Original

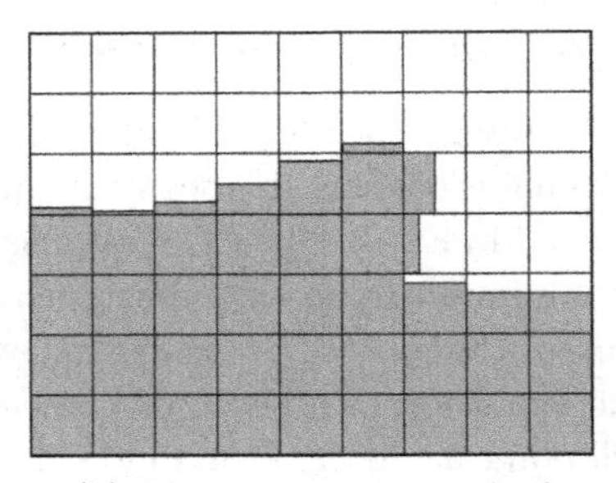

(b) Donor-acceptor method

Fig. 4.2. Modeling the free surface configuration.

$$\frac{\partial F}{\partial t} + u\frac{\partial F}{\partial x} + v\frac{\partial F}{\partial y} + w\frac{\partial F}{\partial z} = 0 \tag{4.2}$$

Eq. (4.2) is thought to indicate a fractional volume of fluid, provided that $F = 0$ and $F = 1$ represent gas and liquid phases, respectively. However, Eq. (4.2) only means that physical quantities are transported by local velocities u, v and w, and F is merely a flag to distinguish between gas and liquid phases. Equation (4.2) basically represents the existence of a free surface when $0 < F < 1$, and F in Eq. (4.2) should not be regarded as the fractional volume of fluid.

To consider F as the fractional volume of fluid, Eq. (4.2) should be combined with the continuity equation, yielding Eq. (4.3).

$$\frac{\partial F}{\partial t} + \frac{\partial Fu}{\partial x} + \frac{\partial Fv}{\partial y} + \frac{\partial Fw}{\partial z} = 0 \tag{4.3}$$

Eq. (4.3) can be used throughout the whole computational domain, and the function F (VOF function) is defined as the fractional volume of the cell occupied by fluid; namely, $F = 1$ corresponds to a cell full of fluid (full cell), while $F = 0$ indicates a cell with no fluid (empty cell). A cell with $0 < F < 1$ must contain a free surface (surface cell), as shown in Fig. 4.2. It should be noted that the advection terms in Eq. (4.3), terms two through four, require highly accurate calculations to prevent the free surface from diffusing. Therefore, a donor-acceptor method,[12] which is described in §4.2.3, is employed with the VOF method.

In summary, the VOF method provides a simple and efficient scheme to capture the free surface with minimum storage requirements, and problems associated with intersecting surfaces can be avoided using the VOF technique. The method is readily extendable to a three-dimensional field without huge storage requirements, as compared with the marker particle method.

4.2.2. *Classification of a cell*

As explained in §4.2.1, a cell is classified as a full, surface or empty cell by the value of the VOF function. However, the direction of the free surface cannot be determined uniquely by using only the value of the VOF function. It is, therefore, necessary to estimate the orientation of the free surface within the surface cell. An example of a method to determine the free surface orientation is explained here.

In the original VOF method, cell types are defined as one of eight conditions in a three-dimensional field: a full cell, a surface cell with the free surface oriented in one of two directions for each of the x-, y-, and z-axes, or an empty cell. The classification of cells by flag NF is shown in Table 4.1 and Fig. 4.3.

To properly impose boundary conditions on the free surface, a surface cell with $0 < F < 1$ has to adjoin both a full cell and an empty cell. To satisfy the above requirements, a surface cell is determined by the following procedure:

(1) All computational cells are assumed to be full cells temporally.
(2) If the VOF function F for a cell is zero, the cell becomes an empty cell.
(3) A cell adjacent to empty cells is regarded as a surface cell.
(4) A surface cell which does not adjoin a full cell becomes an empty cell.
(5) Find surface cells which are not adjacent to both a full cell and an empty cell.
(6) The cells searched in (5) are changed to empty cells.
(7) Repeat the procedure from (3) to (5) until (6) is no longer implemented.

Thus, each cell can be classified as a full cell, an empty cell or a surface cell by the above-mentioned procedure. For the exceptions illustrated in Fig. 4.4, the cell will be correspondingly labeled a fluid cell or an empty cell.

Next, the orientation of each surface cell should be estimated. There are several methods to determine the direction of the free surface in a surface cell. In the simplest method, the orientation of a surface cell can be determined by the adjacent full cells, located in the direction of the maximum fluid amount around the surface cell.

Table 4.1 Classification of a cell

Flag NF	Cell type	Remarks
0	full cell	Full cell occupied fluid
1	surface cell	Free surface is perpendicular to the x-axis and the fluid exists on the negative x-axis side of the cell
2	surface cell	Free surface is perpendicular to the x-axis and the fluid exists on the positive x-axis side of the cell
3	surface cell	Free surface is perpendicular to the y-axis and the fluid exists on the negative y-axis side of the cell
4	surface cell	Free surface is perpendicular to the y-axis and the fluid exists on the positive y-axis side of the cell
5	surface cell	Free surface is perpendicular to the z-axis and the fluid exists on the negative z-axis side of the cell
6	surface cell	Free surface is perpendicular to the z-axis and the fluid exists on the positive z-axis side of the cell
8	surface cell	Cell with no fluid

4.2.3. *Donor-acceptor method*

To capture a highly accurate free surface configuration, Eq. (4.3) needs to be devised in terms of the VOF function. A donor-acceptor method is, therefore, adopted in the original VOF method to compute the advection terms in Eq. (4.3). In the donor-acceptor method, the orientation of the free surface dictates whether an upwind or downwind scheme is selected, which preserves a sharp interface. The maximum advection quantity of the VOF function is also assumed to be the fluid volume contained in a donor cell.

The relation between the free surface orientation and the direction of flux of the VOF function across the cell face is either parallel or perpendicular, because the interface orientation of a surface cell is assumed to coincide with either axis direction. Thus, the flux quantities of the VOF function across cell faces are calculated based on the concept of the donor-acceptor method shown in Fig. 4.5.

As shown in Fig. 4.5(a), when the free surface of the donor cell (the left cell) is perpendicular to the cell face, the value of the VOF function F in

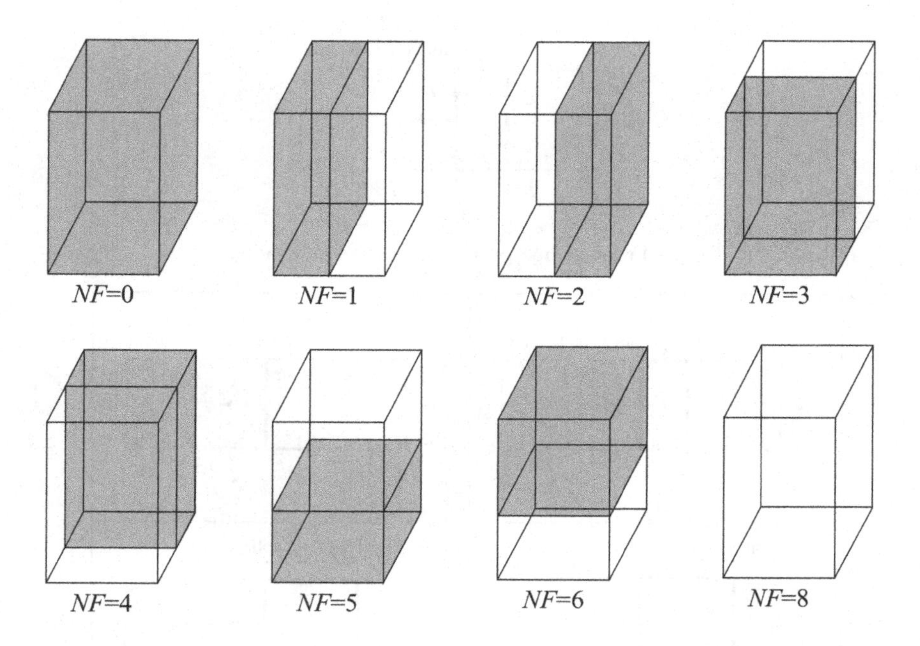

Fig. 4.3. Definition of flag NF.

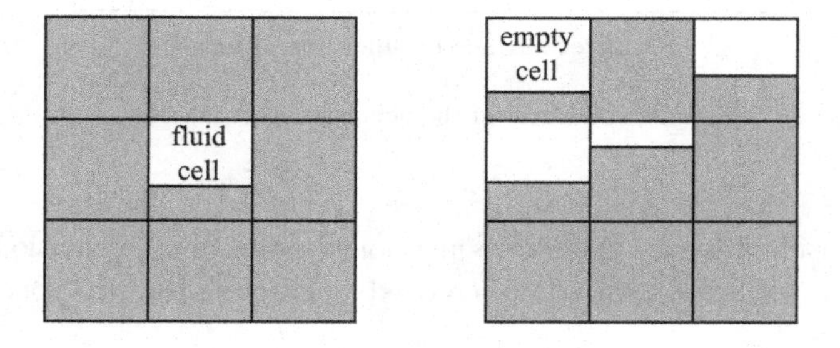

Fig. 4.4. Exceptions in cell classification.

the donor cell on the windward side is used as F on the cell face. On the other hand, when the free surface is parallel to the cell face as shown in Fig. 4.5(b), the value of the VOF function F on the cell face is regarded as F in the acceptor cell on the leeward side. However, it is should be noted that situations may occur when there is not enough fluid or voids in a donor cell, as depicted in Figs. 4.5(b)(3)–(b)(4).

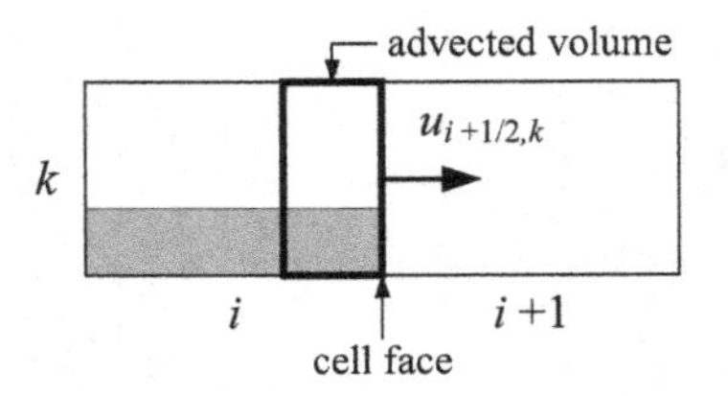

(a) Free surface perpendicular to cell face

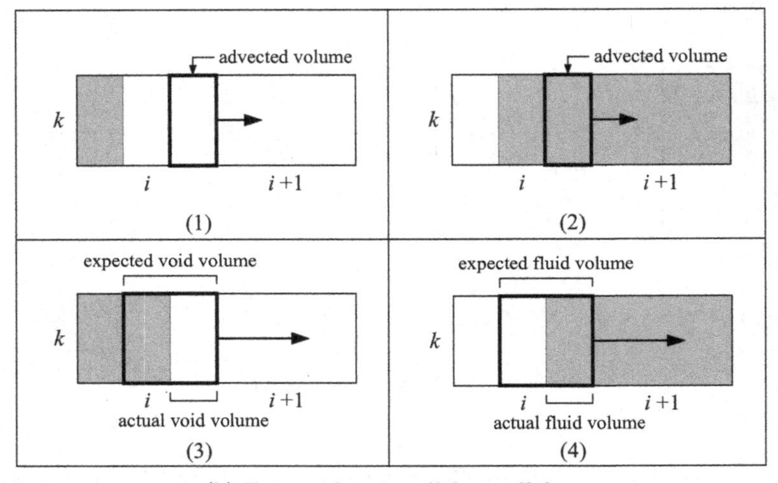

(b) Free surface parallel to cell face

Fig. 4.5. Concept of the donor-acceptor method.

In consideration of the above-mentioned advection calculation procedure of the VOF function, the discretized equation of Eq. (4.3) becomes as follows:

$$\begin{aligned} F_{i,k}^{n+1} = F_{i,k}^{n} &+ \frac{1}{\Delta x_i}(FLFU_{i+1,j,k} - FLFU_{i,j,k}) \\ &+ \frac{1}{\Delta y_j}(FLFV_{i,j+1,k} - FLFV_{i,j,k}) \\ &+ \frac{1}{\Delta z_k}(FLFW_{i,j,k+1} - FLFW_{i,j,k}) \end{aligned} \tag{4.4}$$

where Δx_i, Δy_j, and Δz_k are the respective mesh sizes in the x-, y-, and z-axes. $FLFU$, $FLFV$, and $FLFW$ are inflow or outflow quantities across a cell face for the respective axes, as indicated in the following equations.

$$\begin{aligned} FLFU_{i,j,k} &= -(\Delta t u F)_{i,j,k} \\ &= -\text{sign}(C)\min(F_{AD}\,|C| + CFX, F_D \Delta x_D) \end{aligned} \tag{4.5}$$

where

$$\begin{aligned} CFX &= \max\left[(1 - F_{AD})\,|C| - (F_{DM} - F_D)\,\Delta x_D, 0\right] \\ C &= \Delta t u_{i,j,k} \\ F_{AD} &= \begin{cases} F_D : \text{interface perpendicular to the free surface in donor cell} \\ F_A : \text{other cases} \end{cases} \end{aligned}$$

$$\begin{aligned} FLFV_{i,j,k} &= -(\Delta t v F)_{i,j,k} \\ &= -\text{sign}(C)\min(F_{AD}\,|C| + CFY, F_D \Delta y_D) \end{aligned} \tag{4.6}$$

where

$$\begin{aligned} CFY &= \max\left[(1 - F_{AD})\,|C| - (F_{DM} - F_D)\,\Delta y_D, 0\right] \\ C &= \Delta t v_{i,j,k} \\ F_{AD} &= \begin{cases} F_D : \text{interface perpendicular to the free surface in donor cell} \\ F_A : \text{other cases} \end{cases} \end{aligned}$$

$$\begin{aligned} FLFW_{i,j,k} &= -(\Delta t w F)_{i,j,k} \\ &= -\text{sign}(C)\min(F_{AD}\,|C| + CFZ, F_D \Delta z_D) \end{aligned} \tag{4.7}$$

where

$$\begin{aligned} CFZ &= \max\left[(1 - F_{AD})\,|C| - (F_{DM} - F_D)\,\Delta z_D, 0\right] \\ C &= \Delta t w_{i,j,k} \\ F_{AD} &= \begin{cases} F_D : \text{interface perpendicular to the free surface in donor cell} \\ F_A : \text{other cases} \end{cases} \end{aligned}$$

The function "min" in Eqs. (4.5)–(4.7) prevents the excess accumulation of flux from the donor cell with a value F that is larger than the fluid volume in the donor cell, while the function "max" accounts for additional flux from the VOF function (CFX, CFY, and CFZ) if the amount of the void $(1 - F)$ to be fluxed exceeds the amount of void available in the donor cell.

4.2.4. *Free surface boundary condition*

Two kinds of kinematic and dynamic boundary conditions have to be imposed on the free surface as described below.

(a) Kinematic boundary condition

Solving the advection equation of the VOF function (Eq. (4.3) or Eq. (4.4)) automatically satisfies the kinematic boundary condition for the free surface. Therefore, an additional boundary condition is not required on the free surface.

(b) Dynamic boundary condition

Two dynamic free surface boundary conditions, namely the normal and tangential stress conditions, are required to perform the time-marching computations of the free surface motion.

The normal stress dynamic boundary condition is imposed during the implicit calculation of pressure, ensuring that the pressure on the free surface coincides with atmospheric pressure p_a ($p_a = 0$ in gauge pressure). As a simple example, the pressure in a surface cell can be estimated by linear interpolation or extrapolation between the pressure on the free surface, p_a, and the pressure of the adjacent full cell in the direction indicated by flag NF.

Velocities on empty cells or surface cells around the free surface, which are required to solve the momentum equation, can be evaluated by imposing the tangential stress boundary condition on the free surface. For example, the simple tangential boundary condition can be imposed by setting the velocity on the surface cell interface, which is not adjacent to a full cell, equal to the velocity on the full cell located on the opposite side of the normal direction of the free surface. Further, the velocities on the interfaces between the surface and empty cells are determined such that they satisfy the continuity equation. Therefore, even if the surface cell changes to a full cell at the next time step, the full cell is still able to satisfy the continuity equation.

4.3. Numerical Wave Flume CADMAS-SURF

The numerical wave flume CADMAS-SURF (SUper Roller Flume for Computer Aided Design of MAritime Structure), which is based on the VOF method, is introduced in this section.[4–6]

4.3.1. *Concepts of CADMAS-SURF*

CADMAS-SURF was developed to investigate complicated wave deformation such as wave breaking and wave overtopping and to examine maritime structure design under wave action. The concepts behind CADMAS-SURF model are summarized in Tables 4.2–4.4.

4.3.2. *Governing equation*

The governing equation used in CADMAS-SURF consists of the continuity equation (Eq. (4.8)) and the Reynolds equations (Eqs. (4.9)–(4.11); see §2.2 for the RANS model) based on a porous-body model.[20]

$$\frac{\partial \gamma_x u}{\partial x} + \frac{\partial \gamma_y v}{\partial y} + \frac{\partial \gamma_z w}{\partial z} = S_\rho \tag{4.8}$$

$$\begin{aligned} \lambda_v \frac{\partial u}{\partial t} + \frac{\partial \lambda_x uu}{\partial x} + \frac{\partial \lambda_y vu}{\partial y} + \frac{\partial \lambda_z wu}{\partial z} = \\ -\frac{\gamma_v}{\rho}\frac{\partial p}{\partial x} + \frac{\partial}{\partial x}\left\{\gamma_x \nu_e \left(2\frac{\partial u}{\partial x}\right)\right\} + \frac{\partial}{\partial y}\left\{\gamma_y \nu_e \left(\frac{\partial u}{\partial y} + \frac{\partial v}{\partial x}\right)\right\} \\ + \frac{\partial}{\partial z}\left\{\gamma_z \nu_e \left(\frac{\partial u}{\partial z} + \frac{\partial w}{\partial x}\right)\right\} - \gamma_v D_x u - R_x + \gamma_v S_u \end{aligned} \tag{4.9}$$

$$\begin{aligned} \lambda_v \frac{\partial v}{\partial t} + \frac{\partial \lambda_x uv}{\partial x} + \frac{\partial \lambda_y vv}{\partial y} + \frac{\partial \lambda_z wv}{\partial z} = \\ -\frac{\gamma_v}{\rho}\frac{\partial p}{\partial y} + \frac{\partial}{\partial x}\left\{\gamma_x \nu_e \left(\frac{\partial v}{\partial x} + \frac{\partial u}{\partial y}\right)\right\} + \frac{\partial}{\partial y}\left\{\gamma_y \nu_e \left(2\frac{\partial v}{\partial y}\right)\right\} \\ + \frac{\partial}{\partial z}\left\{\gamma_z \nu_e \left(\frac{\partial v}{\partial z} + \frac{\partial w}{\partial x}\right)\right\} - \gamma_v D_y v - R_y + \gamma_v S_v \end{aligned} \tag{4.10}$$

$$\begin{aligned} \lambda_v \frac{\partial w}{\partial t} + \frac{\partial \lambda_x uw}{\partial x} + \frac{\partial \lambda_y vw}{\partial y} + \frac{\partial \lambda_z ww}{\partial z} = \\ -\frac{\gamma_v}{\rho}\frac{\partial p}{\partial z} + \frac{\partial}{\partial x}\left\{\gamma_x \nu_e \left(\frac{\partial w}{\partial x} + \frac{\partial u}{\partial z}\right)\right\} + \frac{\partial}{\partial y}\left\{\gamma_y \nu_e \left(\frac{\partial w}{\partial y} + \frac{\partial v}{\partial z}\right)\right\} \\ + \frac{\partial}{\partial z}\left\{\gamma_z \nu_e \left(2\frac{\partial w}{\partial z}\right)\right\} - \gamma_v D_z w - R_z + \gamma_v S_w - \frac{\gamma_v \rho^* g}{\rho} \end{aligned} \tag{4.11}$$

where t is the time, the x-, y-, and z-axes are in the Cartesian coordinate

Table 4.2 Features of CADMAS-SURF

Item	Specifications
Applicable numerical analysis	Incompressible flow with multiple free-surfaces
Governing equation	Continuity equation and Navier-Stokes (Reynolds) equation based on porous body model
Coordinate system	Cartesian coordinates
Free surface model	VOF method
Turbulence model	High Reynolds number k-ε two equation turbulence model
Wave generation model	• Wave generation boundary • Wave generation source
Wave generation function	• Fifth-order Stokes wave • Third-order cnoidal wave • Stream function method B • Piston-type wave generator (only 2-D model) • Flap-type wave generator (only 2-D model) • Generation of arbitrary wave with matrix data
Transport of scalar variable	Possible to add arbitrary advection-diffusion equations of scalar variables
Open boundary condition	• Sommerfeld radiation condition • Absorbing sponge zone
Other boundary conditions	• Possible to set obstacles at arbitrary locations • Possible to set boundary conditions on arbitrary positions of obstacle (structure) face • Possible to select boundary conditions such as slip, non-slip, fixed velocity, logarithmic law (smooth and rough surface) and free (in/outflow) conditions using input data

system, u, v, and w are the velocity components in the respective x-, y- and z-axes, ρ is the density, p is the pressure, ν_e is the summation ($\nu_e = \nu + \nu_t$) of molecular viscosity ν and eddy viscosity ν_t, g is the gravitational acceleration, γ_v is the porosity, and γ_x, γ_y, and γ_z are the areal porosities in their respective directions. In addition, D_x, D_y, and D_z are the parameters

Table 4.3 Numerical scheme and algorithm of CADMAS-SURF

Item	Specifications
Discretization	• Finite difference method with staggered mesh • Shape approximation of obstacles with porous body model
Time-integration	Euler method, SMAC method
Numerical scheme for advection term of the equation of motion	Selectable from the following schemes (1) First-order upwind difference scheme (2) Second-order central difference scheme (3) Donor scheme (hybrid of (1) and (2)) (4) QUICK scheme
Treatment of velocity on surface cell	• Extrapolation (extrapolation from velocities inside the fluid) • Zero-gradient (set the same velocity as one on neighbor full cell)
Determination of free surface direction	NASA-VOF3D method
Treatment of air bubble and water droplet	Timer-Door method
Matrix solver	MILU-BiCGSTAB method
Control of time increment	• FIxed (Input) • Automatic (Calculated based on CFL condition)

SMAC : Simplified Marker And Cell
QUICK : Quadratic Upstream Interpolation for Convective Kinematics
MILU-BiCGSTAB : Modified Incomplete Lower-Upper factorization - Bi-Conjugate Gradient STABilized
NASA : National Aeronautics and Space Administration
CFL : Courant-Friedrichs-Lewy

used for the absorbing sponge zone in their respective directions, and S_ρ, S_u, S_v, and S_w are wave generation source terms for the continuity equation, with the equations of motion in the respective directions of x, y, and z, which will be explained in §4.3.3(c).

The coefficients λ_v, λ_x, λ_y and λ_z are denoted as follows. The second terms on the right hand sides of Eq. (4.12), in which C_M is an inertia coefficient, express how the inertia forces are affected by the structure.

$$\left.\begin{aligned}\lambda_v &= \gamma_v + (1-\gamma_v)C_M\\ \lambda_x &= \gamma_x + (1-\gamma_x)C_M\\ \lambda_y &= \gamma_y + (1-\gamma_y)C_M\\ \lambda_z &= \gamma_z + (1-\gamma_z)C_M\end{aligned}\right\} \tag{4.12}$$

Table 4.4 Data-output and visualization of CADMAS-SURF

Item	Specifications
Time-series output data	water surface elevation, velocities, pressure, VOF function and the other scalar quantities at specified locations
Visualization	velocity vector, isogram of variable, free surface, coloring of water region
Animation creation	Visualization data (bmp format) at a fixed time increment

The resistance force R_x, R_y, and R_z from the porous body is modeled as follows:

$$R_x = \frac{1}{2}\frac{C_D}{\Delta x}(1-\gamma_x)\,u\sqrt{u^2+v^2+w^2} \tag{4.13}$$

$$R_y = \frac{1}{2}\frac{C_D}{\Delta y}(1-\gamma_y)\,v\sqrt{u^2+v^2+w^2} \tag{4.14}$$

$$R_z = \frac{1}{2}\frac{C_D}{\Delta z}(1-\gamma_z)\,w\sqrt{u^2+v^2+w^2} \tag{4.15}$$

where C_D is the drag coefficient, and Δx, Δy, and Δz are the mesh sizes in the x-, y-, and z-directions respectively.

The VOF method is adopted in CADMAS-SURF to analyze the time evolution of the free surface configuration. Based on the porous body model, the advection equation of the VOF function F is expressed as follows:

$$\gamma_v\frac{\partial F}{\partial t}+\frac{\partial \gamma_x uF}{\partial x}+\frac{\partial \gamma_y vF}{\partial y}+\frac{\partial \gamma_z wF}{\partial z}=S_F \tag{4.16}$$

where S_F is the source term required for the wave generation source method.

As a turbulence model, the k-ε two-equation model for high Reynolds numbers can be employed in CADMAS-SURF, which is widely used in many research fields (see §2.2 for the details of the k-ε model). The governing equations are the advection-diffusion equations for the turbulence energy k and the turbulence energy dissipation ε, as shown in Eqs. (4.17) and (4.18).

$$\gamma_v\frac{\partial k}{\partial t}+\frac{\partial \gamma_x uk}{\partial x}+\frac{\partial \gamma_y vk}{\partial y}+\frac{\partial \gamma_z wk}{\partial z}=\frac{\partial}{\partial x}\left\{\gamma_x\nu_k\left(\frac{\partial k}{\partial x}\right)\right\} \tag{4.17}$$
$$+\frac{\partial}{\partial y}\left\{\gamma_y\nu_k\left(\frac{\partial k}{\partial y}\right)\right\}+\frac{\partial}{\partial z}\left\{\gamma_z\nu_k\left(\frac{\partial k}{\partial z}\right)\right\}+\gamma_v(G_S+G_T-\varepsilon)$$

$$\gamma_v \frac{\partial \varepsilon}{\partial t} + \frac{\partial \gamma_x u \varepsilon}{\partial x} + \frac{\partial \gamma_y v \varepsilon}{\partial y} + \frac{\partial \gamma_z w \varepsilon}{\partial z} = \frac{\partial}{\partial x}\left\{\gamma_x \nu_\varepsilon \left(\frac{\partial \varepsilon}{\partial x}\right)\right\} + \frac{\partial}{\partial y}\left\{\gamma_y \nu_\varepsilon \left(\frac{\partial \varepsilon}{\partial y}\right)\right\} + \frac{\partial}{\partial z}\left\{\gamma_z \nu_\varepsilon \left(\frac{\partial \varepsilon}{\partial z}\right)\right\} + \gamma_v \left\{C_1 \frac{\varepsilon}{k}(G_S + G_T)(1 + C_3 R_f) - C_2 \frac{\varepsilon^2}{k}\right\} \tag{4.18}$$

where k and ε are defined in the following equations.

$$k = \frac{1}{2}\left(u'^2 + v'^2 + w'^2\right) \tag{4.19}$$

$$\varepsilon = 2\nu \left\{\left(\frac{\partial u'}{\partial x}\right)^2 + \left(\frac{\partial v'}{\partial y}\right)^2 + \left(\frac{\partial w'}{\partial z}\right)^2\right\} + \nu \left\{\left(\frac{\partial u'}{\partial y} + \frac{\partial v'}{\partial x}\right)^2 + \left(\frac{\partial v'}{\partial z} + \frac{\partial w'}{\partial y}\right)^2 + \left(\frac{\partial w'}{\partial x} + \frac{\partial u'}{\partial z}\right)^2\right\} \tag{4.20}$$

where u', v', and w' are velocity fluctuations in the respective directions of x, y, and z.

The eddy viscosity and diffusion coefficients of the above equations are expressed as follows:

$$G_S = 2\nu_t \left\{\left(\frac{\partial u}{\partial x}\right)^2 + \left(\frac{\partial v}{\partial y}\right)^2 + \left(\frac{\partial w}{\partial z}\right)^2\right\} + \nu_t \left\{\left(\frac{\partial u}{\partial y} + \frac{\partial v}{\partial x}\right)^2 + \left(\frac{\partial v}{\partial z} + \frac{\partial w}{\partial y}\right)^2 + \left(\frac{\partial w}{\partial x} + \frac{\partial u}{\partial z}\right)^2\right\} \tag{4.21}$$

$$G_T = -\frac{\nu_t}{\rho \sigma_t}\left(g \frac{\partial \rho^*}{\partial z}\right) \tag{4.22}$$

$$R_f = \frac{-G_T}{G_S + G_T} \tag{4.23}$$

$$\nu_t = C_\mu \frac{k^2}{\varepsilon} \tag{4.24}$$

$$\nu_k = \nu + \frac{\nu_t}{\sigma_k} \tag{4.25}$$

$$\nu_\varepsilon = \nu + \frac{\nu_t}{\sigma_\varepsilon} \tag{4.26}$$

where the empirical coefficients in the above equations are often assigned values of $C_\mu = 0.09$, $\sigma_k = 1.00$, $\sigma_\varepsilon = 1.30$, $C_1 = 1.44$, $C_2 = 1.92$, and $C_3 = 0.0$.

It should be noted that the pressure p in the momentum equations (Eqs. (4.9)–(4.11) has be replaced with $p + 2/3\rho k$ when a turbulence model is used.

4.3.3. *Wave generation model*

(a) Wave generation function

Six types of waves can be selected in CADMAS-SURF, as shown below and in Table 4.2.

- Fifth-order Stokes wave
- Third-order cnoidal wave
- Stream function method B
- Piston-type wave generator (only 2-D model)
- Flap-type wave generator (only 2-D model)
- Generation of arbitrary wave with matrix data

The computed water surface elevation $\eta_S(t)$ may not coincide with the theoretical water surface elevation $\eta_0(t)$ based on the above theories for cases where reflected waves are generated. To overcome this problem, the input horizontal velocity $U(z,t)$ in the computations is adjusted so that the inflow flux will balance the outflow flux at the wave generation line by introducing Eq. (4.27).

$$U(z,t) = U_0(z^*,t) \cdot \left(\frac{\eta_0 + h}{\eta_s + h} \right) \tag{4.27}$$

$$z^* = \frac{\eta_0 + h}{\eta_s + h}(z + h) - h \tag{4.28}$$

where h is the still water depth.

(b) Wave generation boundary method

At the specified boundaries, the wave generation boundary method sets the water surface elevation and horizontal and vertical velocities based on the above-mentioned wave theories. The boundary condition of a zero gradient for both the VOF function F and the pressure p should be imposed to prevent excess inflow or outflow at the wave generation boundaries. However, this method has a drawback in that wave generation cannot be implemented properly when incident waves and reflected waves interact in the computational domain.

General interpretations of the boundary conditions for water wave computations are presented in §3.2.

(c) Wave generation source method

In the wave generation source method, waves can be generated by introducing a source or sink in the computational domain, so that reflected waves do not influence the incident wave used as input in the computations. The source terms in Eqs. (4.8)–(4.11) and Eq. (4.16) are concretely shown below:

$$S_\rho = q(z,t) \tag{4.29}$$

$$S_u = uq(z,t) \tag{4.30}$$

$$S_v = vq(z,t) \tag{4.31}$$

$$S_w = wq(z,t) + \frac{\nu}{3}\frac{\partial q(z,t)}{\partial z} \tag{4.32}$$

$$S_F = Fq(z,t) \tag{4.33}$$

where $q(z,t)$ is the wave generation source and is expressed with mesh size Δx_s at the wave source line (for the case of wave generation toward the x-direction) in the following equation.

$$q(z,t) = 2U\left(z,t\right)/\Delta x_s \tag{4.34}$$

The coefficient "2" on the right-hand side of Eq. (4.34) corresponds to two propagating waves on both the left and right sides of the wave generation source.

4.3.4. *Open boundary treatment*

An open boundary treatment is one of the important problems in computations demanding high accuracy. CADMAS-SURF has two types of open boundary treatments to maintain stable calculations, namely the Sommerfeld radiation condition and the absorbing sponge zone. See also general descriptions about the boundary condition in §3.2.2 and §3.2.3.

(a) Sommerfeld radiation condition

The Sommerfeld radiation boundary condition is as follows:

$$\frac{\partial f}{\partial t} + C\frac{\partial f}{\partial x} = 0 \qquad \text{or} \qquad \frac{\partial f}{\partial t} + C\frac{\partial f}{\partial y} = 0 \tag{4.35}$$

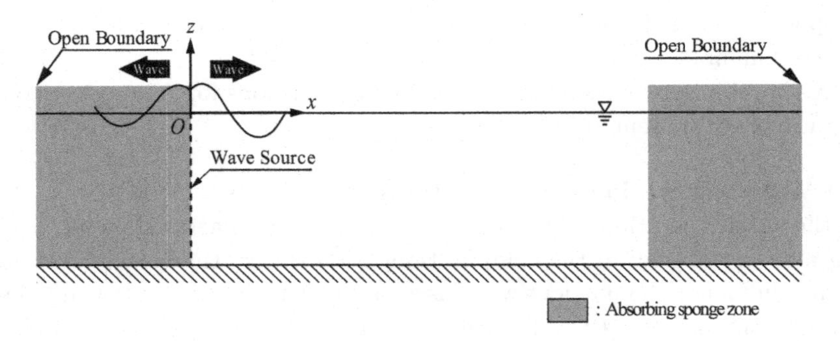

Fig. 4.6. Definition sketch of a numerical wave tank.

where f is a physical variable such as velocity or VOF function, and C is the wave celerity. C can be set based on small amplitude wave theory, but further studies are still required to apply this boundary condition to strongly nonlinear wave fields, whether they are regular or irregular.

(b) Absorbing sponge zone

An absorbing sponge zone (layer) is the zone where the wave height decreases due to forcibly attenuating wave energy. The zone has a length of about one to three wavelengths. The advantage of the sponge zone is that it enables stable calculations under any wave action, although the zone is an extra computational domain and requires long computational time. Parameters related to the energy dissipation D_x, D_y, and D_z in Eqs. (4.9)–(4.11) are set to be proportional to velocity in their respective directions, as shown in the following equations.[7]

$$D_x = D_y = \theta_{xy}\sqrt{\frac{g}{h}}(N+1)\left(\frac{\max\left(|x-x_0|, |y-y_0|\right)}{l}\right)^N \tag{4.36}$$

$$D_z = \theta_z\sqrt{\frac{g}{h}}(N+1)\left(\frac{\max\left(|x-x_0|, |y-y_0|\right)}{l}\right)^N \tag{4.37}$$

where h is the still water depth, l and x_0 or y_0 are the width and start positions of the sponge zone, respectively. N is the degree of a distribution function, and θ_x, θ_y, and θ_z are non-dimensional coefficients.

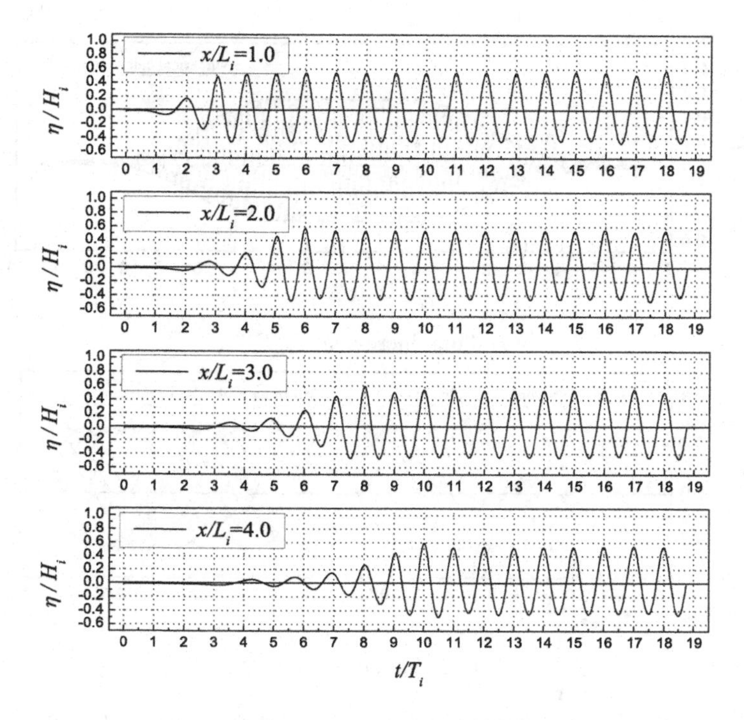

Fig. 4.7. Time variation of water surface profile ($H_i/L_i = 0.04, h/L_i = 0.4$).

4.4. Application of the VOF Method

4.4.1. *Verification of a numerical wave flume*

This subsection verifies the validity and utility of a numerical wave flume, which consists of a wave generation source method and an open boundary treatment due to a sponge zone.[15,16] A definition sketch of a numerical wave flume is shown in Fig. 4.6. The computational domain includes the sponge zone (12 m) and is 16 m and 0.60 m in the respective directions of x and z. The origin of x coincides with the wave generation source. The positive direction of the x-axis is toward the right hand side. The vertical z-axis is positive upward with its origin being the still water level. Mesh sizes in the x- and z-directions, $\Delta x/L_i$ (L_i: wavelength) and $\Delta z/h$, are 1/200 in the analytical domain ($0.0 \le x/L_i \le 4.0$) and 1/40, respectively. The time interval Δt is set constant to be 0.001s so that the CFL condition is always satisfied.

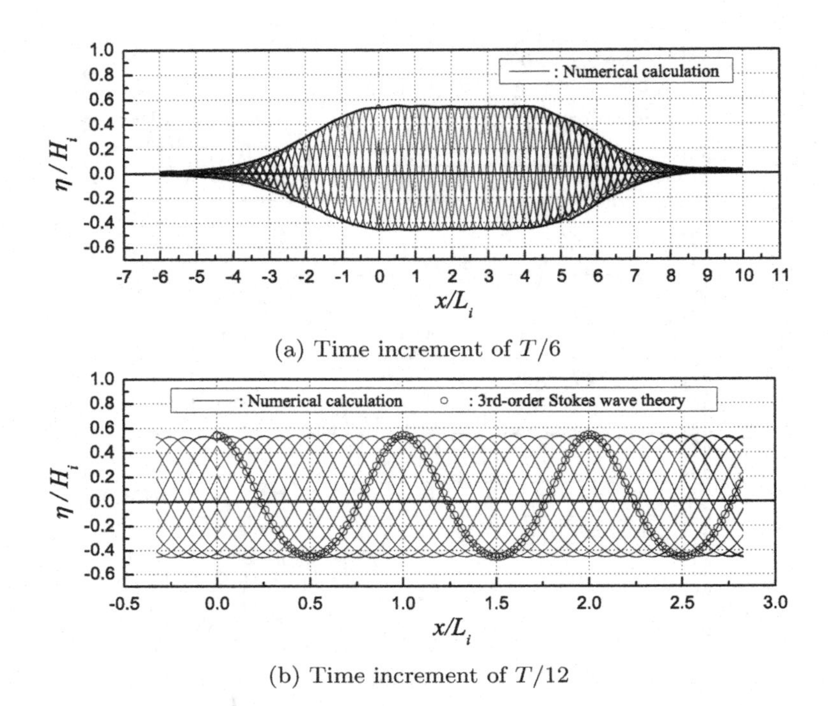

(a) Time increment of $T/6$

(b) Time increment of $T/12$

Fig. 4.8. Spatial profile of the water surface ($H_i/L_i = 0.03, h/L_i = 0.2$).

The time histories of the water surface profile η at $x/L_i = 1.0, 2.0, 3.0$, and 4.0 are shown in Fig. 4.7 with the wave steepness $H_i/L_i = 0.04$ and the relative water depth $h/L_i = 0.4$. It is clear that the computed water surface profile becomes stable and regular from the seventh wave after the start of wave generation.

Figure 4.8 shows the computed spatial profile of the water surface $\eta(x)$ (thin solid line) from the thirteenth to eighteenth wave periods after the start of wave generation. Figures 4.8(a) and (b) illustrate the water surface profiles at a time increment of $T_i/6$ and $T_i/12$, respectively. The corresponding theoretical value of the third-order Stokes wave theory (open circle) is also plotted for comparison. It is found from Fig. 4.8(a) that the spatial profiles of the water surface form a constant envelope curve in the analytical domain ($0.0 \leq x/L_i \leq 4.0$) except for the sponge zone ($x/L_i < 0.0$, $x/L_i > 4.0$). On the other hand, the spatial envelope curve of the water surface is gradually attenuated in the sponge zone, and vanishes near the outmost open boundaries. Moreover, both the theoretical

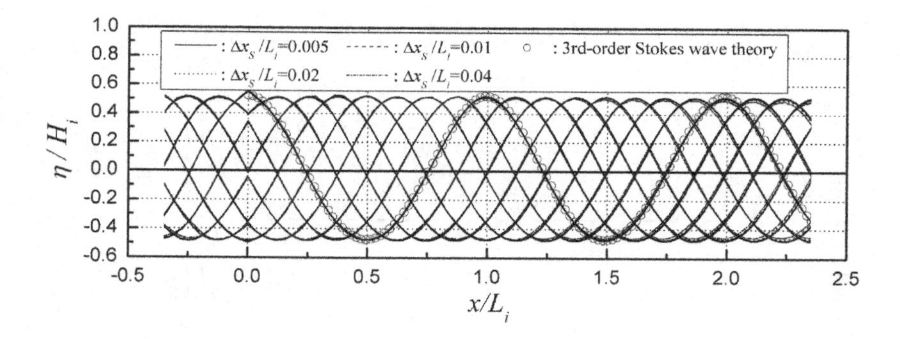

Fig. 4.9. Spatial profile of the water surface ($H_i/L_i = 0.03, h/L_i = 0.4$).

and numerical results agree well, as shown in Fig. 4.8(b). The computed wavelength is 195.5cm, and this is very close to $L_i = 195.42$ cm predicted by the third-order Stokes wave theory.

One should verify whether the mesh size Δx_s at the wave source line has an influence on wave generation, because the wave generation source q^* is given as $q(z,t)/\Delta x_s$, as explained in §4.2.3(c). As can been seen from Fig. 4.9, which indicates the spatial profiles of the water surface for $\Delta x_s/L_i = 0.005$, 0.01, 0.02 and 0.04, the mesh size Δx_s has no influence on the computed water surface profile, and the numerical results are in good agreement with theoretical values.

It is verified that the numerical wave flume can perform the calculations for a stable unsteady condition without significant reflection due to the open boundary treatment. Also, the numerical wave flume can generate arbitrary waves by choosing its corresponding strength q^*.

4.4.2. *Wave breaking on a slope*

In designing coastal structures, it is important to comprehend and predict the wave breaking point and post-breaking wave deformation.

The validity of CADMAS-SURF/3D for wave breaking simulations[2] is verified by comparing the numerical results with the breaker index of Goda.[8,9] Computational domains for slopes of 1/10, 1/20, 1/30, and 1/50 are shown in Fig. 4.10. Figure 4.11 depicts the relation between H_b/h_b and h_b/L_0, in which H_b and h_b are the wave height and water depth at the wave breaking point, and L_0 is the wavelength of the deep water wave. Symbols and solid lines in the figure illustrate the numerical results and

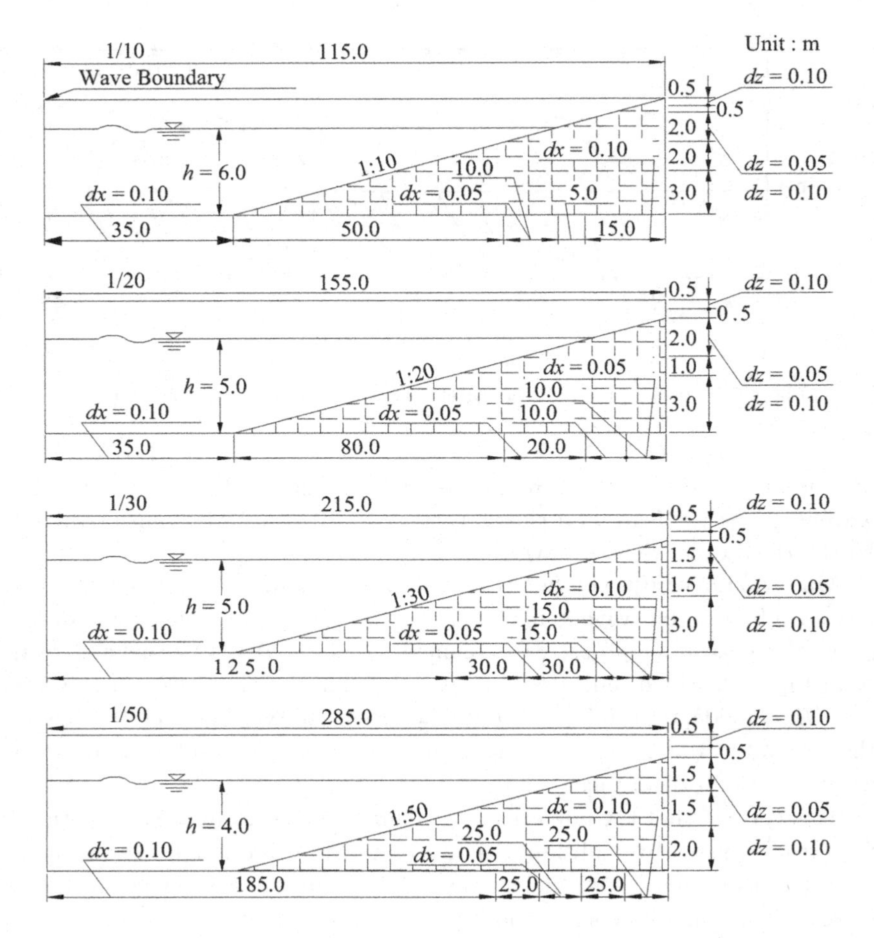

Fig. 4.10. Computational domains for slopes 1/10, 1/20,1/30 and 1/50.

Goda's breaker index, respectively. The numerical results are found to be similar to the empirical breaker indices determined from the laboratory experiments.

Figure 4.12 shows the relationship between the relative water depth h_b/H_0' at the wave breaking point and the ratio H_0'/L_0 of the equivalent wave height and wavelength of the deep water wave. As recognized from the figure, CADMAS-SURF/3D is capable of reproducing the previous experimental data of Bowen *et al.*[3] and Iversen[14] well.

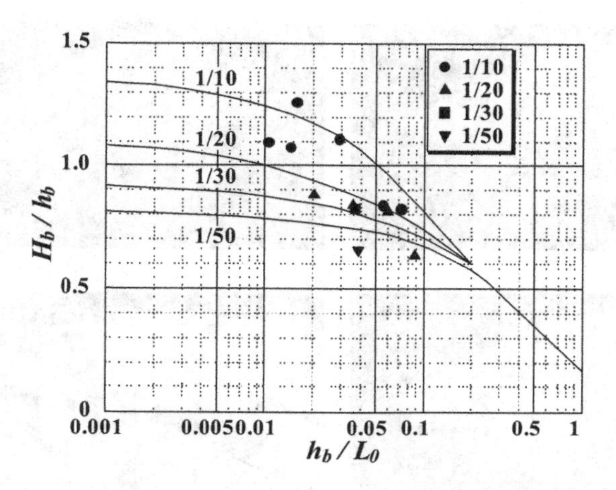

Fig. 4.11. Comparison with Goda's breaker index.

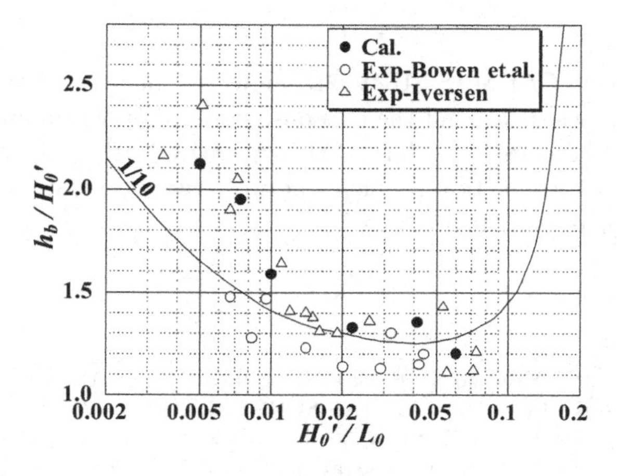

Fig. 4.12. Comparison with previous experimental data.

4.4.3. *Wave and structure interaction*

One example of a wave and structure interaction problem occurs when regions of low atmospheric pressure cause high waves and wave overtopping around caisson breakwaters.[1] Figure 4.13, which shows the numerical results of the interaction between obliquely incident waves and caisson

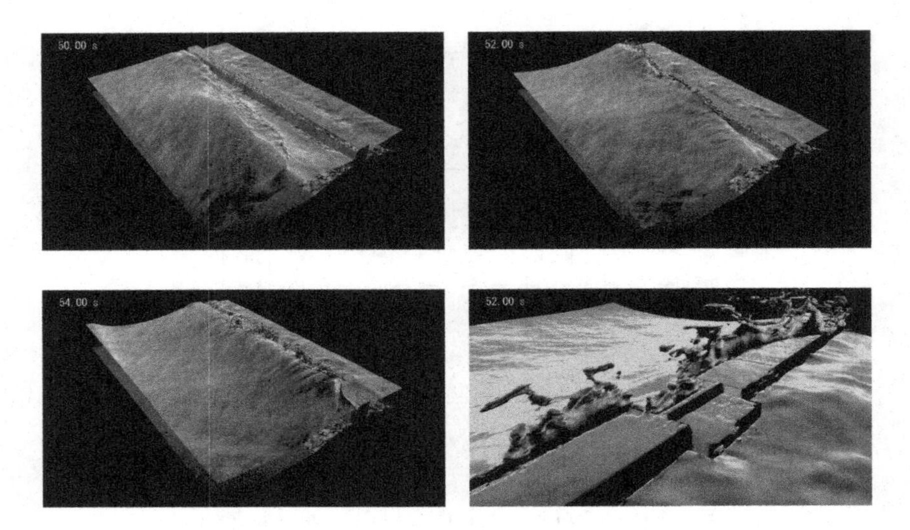

Fig. 4.13. Numerical result of wave deformation around cassion breakwaters.

breakwaters using CADMAS-SURF/3D, indicates that wave breaking and large splash-up occurs around the caissons, resulting in an impulsive pressure acting on the caissons. Thus, CADMAS-SURF/3D would be able to simulate complex interaction problems between waves and structures in a three-dimensional wave field.

References

1. Arikawa T. and Yamano T. (2008): Large-Scale Simulations on Impulsive Wave Pressures by using CADMAS-SURF/3D, *Annual Journal of Coastal Engineering*, JSCE, Vol.55, pp.26-30 (in Japanese).
2. Arikawa T., Yamano T. and Akiyama M. (2007): Advanced Deformation Method for Breaking Waves by Using CADMAS-SURF/3D, *Annual Journal of Coastal Engineering*, JSCE, Vol.54, pp.71-75 (in Japanese).
3. Bowen A.J., Inman D.L. and Simmons V.P. (1968): Wave eSet-Downf and eSet-Upf, *Journal of Geophysical Research*, Vol. 73, No. 8, pp.2569-2577.
4. Coastal Development Institute of Technology (2001): *Research and Development of Numerical Wave Flume "CADMAS-SURF"*, 457p. (in Japanese).
5. Coastal Development Institute of Technology (2008): *CADMAS-SURF Practical Computation Casebook* (in Japanese).
6. Coastal Development Institute of Technology (2010): *Research and Development of Numerical Wave Tank "CADMAS-SURF/3D"* (in Japanese).

7. Cruz E.C. and Isobe M. (1994): Numerical Wave Absorbers for Short and Long Wave Modeling, *Proc. Int. Symp. Waves - Physical and Numerical Modeling*, Vol.2, pp.992-1001.
8. Goda Y. (1970): A Synthesis of Breaker Indices, *Transactions of the Japan Society of Civil Engineers*, Vol. 2(2), pp. 227-230 (in Japanese).
9. Goda Y. (2010): *Random Seas and Design of Maritime Structures*, 3rd Edition, Advanced Series on Ocean Engineering, Vol. 33, World Scientific Publishing Co. Pte. Ltd.
10. Harlow F.H., Amsden A.A. and Nix J.R. (1976): Relativistic Fluid Dynamics Calculations with the Particle-in-Cell Technique, *Journal of Computational Physics*, Vol. 20, pp.119-129.
11. Harlow F.H. and Welch J.E. (1965): Numerical Calculation of Time-Dependent Viscous Incompressible Flow of Fluid with Free Surface, *Physics of Fluids*, Vol. 8, pp.2182-2189.
12. Hirt C.W. and Nichols B.D. (1981): Volume of Fluid (VOF) Method for the Dynamics of Free Boundaries, *Journal of Computational Physics*, Vol.39, pp.201-225.
13. Hirt C.W., Nichols B.D. and Romero J.L. (1975): SOLA-A Numerical Solution Algorithm for Transient Fluid Flows, *Los Alamos Scientific Laboratory Report LA-5852.*
14. Iversen H.W. (1952): Waves and Breakers in Shoaling Water, *Proc. 3rd Int. Conf. on Coastal Engineering*, ASCE, pp.1-12.
15. Kawasaki K. (1999): Numerical Simulation of Breaking and Post-Breaking Wave Deformation Process around a Submerged Breakwater, *Coastal Engineering Journal*, Vol.41, Nos.3 & 4, pp.201-223.
16. Kawasaki K. and Iwata K. (1999): Numerical Analysis of Wave Breaking Due to Submerged Breakwater in Three-Dimensional Wave Field, *Proc. 27th Int. Conf. on Coastal Engineering*, 1998, Vol.1, pp.853-866.
17. Nichols B.D. and Hirt C.W. (1971): Improved Free Surface Boundary Conditions for Numerical Incompressible-Flow Calculations, *Journal of Computational Physics*, Vol. 8, pp.434-448.
18. Nichols B.D. and Hirt C.W. (1973): Calculating Three-Dimensional Free Surface Flows in the Vicinity of Submerged and Exposed Structures, *Journal of Computational Physics*, Vol. 12, pp.234-246.
19. Nichols B.D. and Hirt C.W. (1975): Methods for Calculating Multi-Dimensional, Transient Free Surface Flows Past Bodies, *Los Alamos Scientific Laboratory Technical Report LA-UR-75-1939.*
20. Sakakiyama T. and Kajima R. (1994): Numerical Simulation of Nonlinear Wave Interacting with Permeable Breakwaters, *CRIEPI (Central Research Institute of Electric Power Industry) Research Report*, U93052, pp.1-45 (in Japanese).
21. Welch J.E., Harlow F., Shannon J.P. and Daly B.J. (1966): The MAC Method: A Computing Technique for Solving Viscous, Incompressible, Transient Fluid Flow Problems Involving Free-Surface, *Los Alamos Scientific Laboratory Report LA-3452.*

Chapter 5

CIP Method

Yasunori Watanabe

The Constrained Interpolation Profile (or Cubic Interpolated Pseudo Particle), a so-called CIP method, is a semi-Lagrangian numerical solver for a hyperbolic equation.[24,35] This method explicitly transports a physical quantity and its first derivatives, interpolated over a computing cell by third-order polynomials, along the characteristic curve of the hyperbolic equation. An advantage of this method allows for the computation of the fluid dynamics of moving interfaces, such as compressible fluids around a shock wave, or gas-liquid interfacial flows.

Yabe and his colleagues have proposed many versions of CIP methods to extend this application, improving numerical conservation and accuracy. These versions have been applied to wide-ranging scientific and engineering research fields.

In this chapter, the theoretical background and standard CIP approach are first interpreted in §5.1. A practical technique for CFD and other versions of CIP are explained in §5.2 and 5.3. Major computational techniques for free-surface flow and applications to water wave computations are introduced in §5.4 and 5.5.

5.1. Outline of the CIP Method

The CIP method provides semi-Lagrangian approximations for both the physical quantities and their first spatial derivatives following the advection equations, which is explicitly achieved by quasi-steady transport of spatially interpolated quantities within a computing cell.

First, to understand the fundamental framework of this approach, the simplest one dimensional case is considered here. A one dimensional advection equation for a variable $f(x,t)$ is written as

$$\frac{\partial f}{\partial t} + u\frac{\partial f}{\partial x} = 0 \tag{5.1}$$

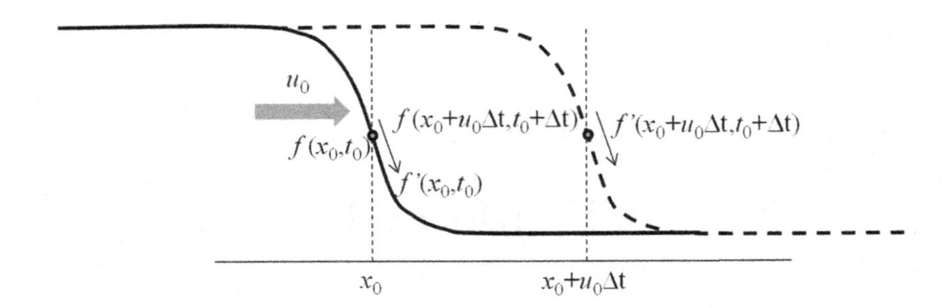

Fig. 5.1. Advection of a quantity f and its first derivative f' transported with velocity u_0. The distributions of f and f' are preserved during advection with CIP.

where $u(x,t)$ is the advection velocity. Differentiating equation (5.1) with respect to x, we have

$$\frac{\partial f'}{\partial t} + u\frac{\partial f'}{\partial x} = -\frac{\partial u}{\partial x}f \tag{5.2}$$

where the superscript $'$ represents the partial derivative, $\partial/\partial x$. Since the right hand side of Eq. (5.2) becomes 0 for the case of constant velocity $\partial u/\partial x = 0$, Eq. (5.2) also describes advection of the first derivative of f; that is, both f and f' are governed by advection Eqs. (5.1) and (5.2) and are transported by velocity u (see Fig. 5.1).

With the CIP method, any variables following the advection equation are spatially interpolated between discrete adjacent grid points i and $i-1$ by Hermite spline polynomials. For example, the distribution form of f between the coordinates x_{i-1} and x_i with grid spacing Δx is described by

$$f(x) = a(x-x_i)^3 + b(x-x_i)^2 + c(x-x_i) + d \tag{5.3}$$

where a through d are local constants to be determined later. Differentiating Eq. (5.3) with respect to x, f' is also interpolated over these grid points

$$f'(x) = 3a(x-x_i)^2 + 2b(x-x_i) + c \tag{5.4}$$

Since f_i and f_i' are known and given at the location $x = x_i$, Eqs. (5.3) and (5.4) provide the relations

$$f(x_i) = d = f_i \tag{5.5}$$

$$f'(x_i) = c = f_i' \tag{5.6}$$

Substituting known values f_{i-1} and f'_{i-1} at $x = x_{i-1} = x_i - \Delta x$ into Eqs. (5.3) and (5.4) as well as the above relationships, we have

$$f(x_{i-1}) = -a\Delta x^3 + b\Delta x^2 - f'_i \Delta x + f_i = f_{i-1} \tag{5.7}$$

$$f'(x_{i-1}) = 3a\Delta x^2 - 2b\Delta x + f'_i = f'_{i-1} \tag{5.8}$$

which determine the constants a and b:

$$a = \frac{f'_i + f'_{i-1}}{\Delta x^2} + 2\frac{f_{i-1} - f_i}{\Delta x^3} \tag{5.9}$$

$$b = 3\frac{f_{i-1} - f_i}{\Delta x^2} + \frac{f'_{i-1} + 2f'_i}{\Delta x} \tag{5.10}$$

Assuming local, steady advection for an infinitesimal duration, Δt, $f(x_0)$ and $f'(x_0)$ at an arbitrary location $x = x_0$ (where $x_{i-1} \le x_0 \le x_i$) are transported at velocity u_0 to the location $x = x_0 + \int_{\Delta t} x_0 dt \equiv x_0 + u_0 \Delta t$ during Δt by following Eqs. (5.1) and (5.2) (see Fig. 5.1). In the same way, we can suppose that $f^{n+1}(x_i)$ and $f'^{n+1}(x_i)$ at $t = t_{n+1} = t_n + \Delta t$ are transported from the upwind location $x = x_i - u_i \Delta t$ at $t = t_n$. Therefore, the advection equations Eqs. (5.1) and (5.2) are solved in a quasi-Lagrangian manner as $f_i^{n+1} = f^n(x_i - u_i\Delta t)$ and $f_i'^{n+1} = f'(x_i - u_i\Delta t)$, which indicates explicit advection of the distribution form of f along the characteristic curve of a hyperbolic equation (see Fig. 5.2). Since $f(x)$ and $f'(x)$ at an arbitrary location between x_{i-1} and x_i are estimated through interpolation by Eqs. (5.3) and (5.4) with the above constants from Eqs. (5.9) and (5.10) with third order accuracy, we finally have

$$f_i^{n+1} = a\xi^3 + b\xi^2 + f_i'^n \xi + f_i^n \tag{5.11}$$

$$f_i'^{n+1} = 3a\xi^2 + 2b\xi + f_i'^n \tag{5.12}$$

where the upwind displacement $\xi = -u_i\Delta$. In this way, f and f' are explicitly updated using those quantities only at the two adjacent grid points $i-1$ and i. This compactness of the CIP computation is especially advantageous when handling fluid motion near a boundary.

5.1.1. *Two-dimensional interpolation*

A multi-dimensional computation is necessary for practical applications representing general fluid motion. In this section, a two-dimensional version of Hermite spline polynomials is used to interpolate variables over computing grids in two dimensional space (x, y) to solve the advection equation for $f(x, y, t)$:

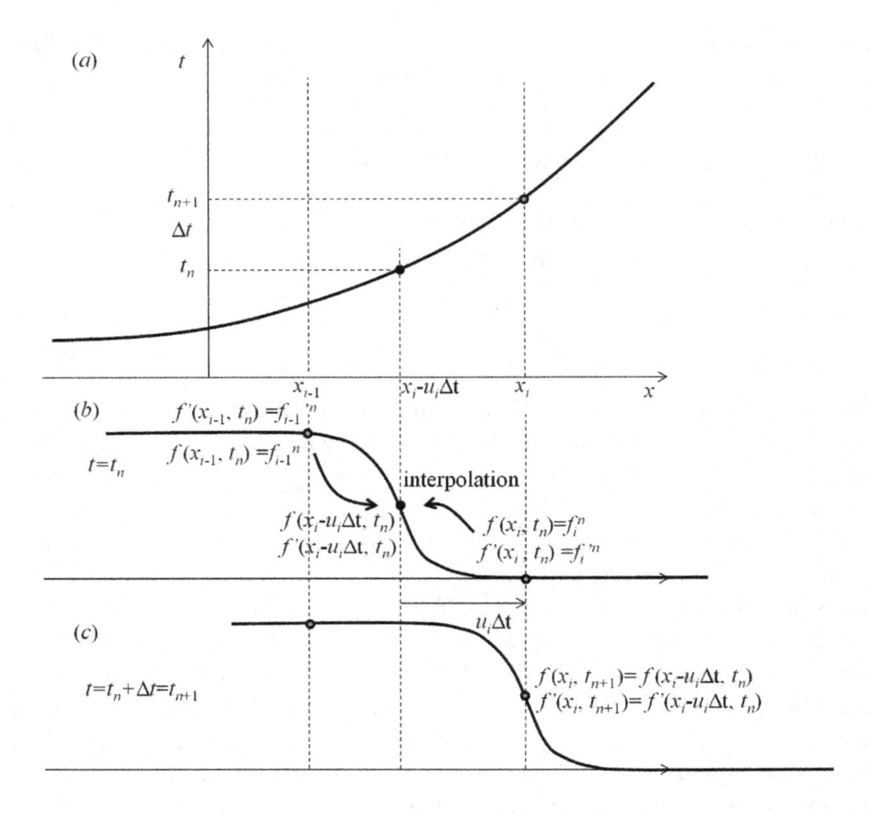

Fig. 5.2. Schematic representation of the CIP approach: (a) a characteristic curve of the advection equation, (b) the distributions of f at $t = t_n$, and (c) $t = t_{n+1} = t_n + \Delta t$. The distribution form of f between x_{i-1} and x_i at $t = t_n$ is approximated by Hermite spline polynomials (b) to be transported to the point (c) along the characteristic curve.

$$\frac{Df}{Dt} = \frac{\partial f}{\partial t} + \boldsymbol{u} \cdot \boldsymbol{\nabla} f = \frac{\partial f}{\partial t} + u\frac{\partial f}{\partial x} + v\frac{\partial f}{\partial y} = 0 \tag{5.13}$$

We will consider planar flow in a Cartesian grid system (i, j) with uniform grid widths Δx and Δy in the i- and j-directions, respectively. Since the quantity $f(x, y, t)$ is transported in the direction of the velocity vector $\boldsymbol{u}(u, v)$ in two dimensional space, $f_{i,j}^{n+1}$ and its spatial derivatives $\boldsymbol{\nabla} f_{i,j}^{n+1}$ at the coordinate $\boldsymbol{x}_{i,j}$ at the $n + 1$ time step are estimated to be identical to $f^n(x_{i,j} - \xi, y_{i,j} - \eta)$ at the upwind location $\boldsymbol{\xi} = (\xi, \eta) = (u_{i,j}\Delta t, v_{i,j}\Delta t)$ from $\boldsymbol{x}_{i,j}$ at the nth time step (see Fig. 5.3 for the case $u < 0$ and $v < 0$):

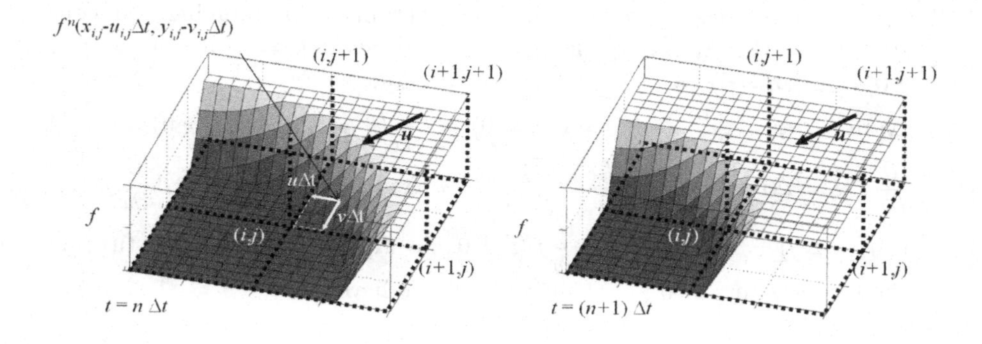

Fig. 5.3. Advection in a two-dimensional grid system; $f^n\left(x_{i,j}-u_{i,j}\Delta t, y_{i,j}-v_{i,j}\Delta t\right)$ at the upwind location is transported to $(x_{i,j}, y_{i,j})$ during Δt for the case of $u<0$ and $v<0$.

$$f_{i,j}^{n+1} = f^n\left(x_{i,j}-\xi, y_{i,j}-\eta\right) \tag{5.14}$$

$$\frac{\partial f_{i,j}^{n+1}}{\partial x} = \frac{\partial f^n\left(x_{i,j}-\xi, y_{i,j}-\eta\right)}{\partial x} \tag{5.15}$$

$$\frac{\partial f_{i,j}^{n+1}}{\partial y} = \frac{\partial f^n\left(x_{i,j}-\xi, y_{i,j}-\eta\right)}{\partial y} \tag{5.16}$$

In two-dimensional CIP, two-dimensional interpolation is performed for the cell containing the upwind coordinate $(x_{i,j}-u_{i,j}\Delta t, y_{i,j}-v_{i,j}\Delta t)$ for the advected flow.

The quantities f and ∇f at an arbitrary location relative to the grid coordinate $\boldsymbol{x}_{i,j}$ where $\boldsymbol{\xi}=(\xi,\eta)=\boldsymbol{x}-\boldsymbol{x}_{i,j}$ are interpolated over all the grid points (i,j), $(i+1,j)$, $(i,j+1)$, and $(i+1,j+1)$ by two dimensional Hermite spline polynomials (see Fig. 5.4):

$$f(\xi,\eta) = \{(a_1\xi + a_2\eta + a_3)\xi + a_4\eta + a_5\}\xi + \{(a_6\eta + a_7\xi + a_8)\eta + a_9\}\eta + a_{10} \tag{5.17}$$

Differentiating the above equation with respect to x and y, the first derivatives to be interpolated are expressed as

$$\frac{\partial f(\xi,\eta)}{\partial x} = (3a_1\xi + 2a_2\eta + 2a_3)\xi + (a_4 + a_7\eta)\eta + a_5 \tag{5.18}$$

$$\frac{\partial f(\xi,\eta)}{\partial y} = (a_2\xi + 2a_7\eta + a_4)\xi + (2a_8 + 3a_6\eta)\eta + a_9 \tag{5.19}$$

The local coefficients a_1 to a_{10} are determined by fulfilling ten constraints of the known quantities at the grids: $f(\xi = 0, \eta = 0) = f_{i,j}$; $f(\xi = \Delta x, \eta = 0) = f_{i+1,j}$; $f(\xi = 0, \eta = \Delta y) = f_{i,j+1}$; $f(\xi = \Delta x, \eta = \Delta y) = f_{i+1,j+1}$; $\frac{\partial}{\partial x} f(\xi = 0, \eta = 0) = \frac{\partial f_{i,j}}{\partial x}$; $\frac{\partial}{\partial y} f(\xi = 0, \eta = 0) = \frac{\partial f_{i,j}}{\partial y}$; $\frac{\partial}{\partial x} f(\xi = \Delta x, \eta = 0) = \frac{\partial f_{i+1,j}}{\partial x}$; $\frac{\partial}{\partial y} f(\xi = \Delta x, \eta = 0) = \frac{\partial f_{i+1,j}}{\partial y}$; $\frac{\partial}{\partial x} f(\xi = 0, \eta = \Delta y) = \frac{\partial f_{i,j+1}}{\partial x}$; and $\frac{\partial}{\partial y} f(\xi = 0, \eta = \Delta y) = \frac{\partial f_{i,j+1}}{\partial y}$. Substituting these constraints into Eqs. (5.17)–(5.19), we have

$$
\begin{aligned}
a_1 &= \frac{1}{\Delta x^3} \left\{ 2\left(f_{i,j} - f_{i+1,j}\right) + \left(\frac{\partial f_{i,j}}{\partial x} + \frac{\partial f_{i+1,j}}{\partial x} \right) \Delta x \right\} \\
a_2 &= \frac{1}{\Delta x^2 \Delta y} \left\{ f_{i,j} - f_{i,j+1} - f_{i+1,j} + f_{i+1,j+1} + \left(\frac{\partial f_{i,j}}{\partial x} - \frac{\partial f_{i,j+1}}{\partial x} \right) \Delta x \right\} \\
a_3 &= \frac{1}{\Delta x^2} \left\{ 3\left(f_{i+1,j} - f_{i,j}\right) - \left(2\frac{\partial f_{i,j}}{\partial x} + \frac{\partial f_{i+1,j}}{\partial x} \right) \Delta x \right\} \\
a_4 &= \frac{1}{\Delta x \Delta y} \left\{ -f_{i,j} + f_{i,j+1} + f_{i+1,j} - f_{i+1,j+1} \right. \\
&\quad \left. + \left(-\frac{\partial f_{i,j}}{\partial x} + \frac{\partial f_{i,j+1}}{\partial x} \right) \Delta x + \left(-\frac{\partial f_{i,j}}{\partial y} + \frac{\partial f_{i+1,j}}{\partial y} \right) \Delta y \right\} \\
a_5 &= \frac{\partial f_{i,j}}{\partial x} \\
a_6 &= \frac{1}{\Delta y^3} \left\{ 2\left(f_{i,j} - f_{i,j+1}\right) + \left(\frac{\partial f_{i,j}}{\partial y} + \frac{\partial f_{i,j+1}}{\partial y} \right) \Delta y \right\} \\
a_7 &= \frac{1}{\Delta x \Delta y^2} \left\{ f_{i,j} - f_{i,j+1} - f_{i+1,j} + f_{i+1,j+1} + \left(\frac{\partial f_{i,j}}{\partial y} - \frac{\partial f_{i+1,j}}{\partial y} \right) \Delta y \right\} \\
a_8 &= \frac{1}{\Delta y^2} \left\{ 3\left(f_{i,j+1} - f_{i,j}\right) - \left(2\frac{\partial f_{i,j}}{\partial y} + \frac{\partial f_{i,j+1}}{\partial y} \right) \Delta y \right\} \\
a_9 &= \frac{\partial f_{i,j}}{\partial y} \\
a_{10} &= f_{i,j}
\end{aligned}
$$

In multi-dimensional CIP, depending on the direction of velocity $\boldsymbol{u}_{i,j}$, we need to choose the upwind cell to perform the interpolation. For the case where $u_{i,j} < 0$ and $v_{i,j} < 0$, the upwind location $\boldsymbol{x} = (x_{i,j} - u_{i,j}\Delta t, y_{i,j} - v_{i,j}\Delta t)$ must be somewhere in the first quadrant cell

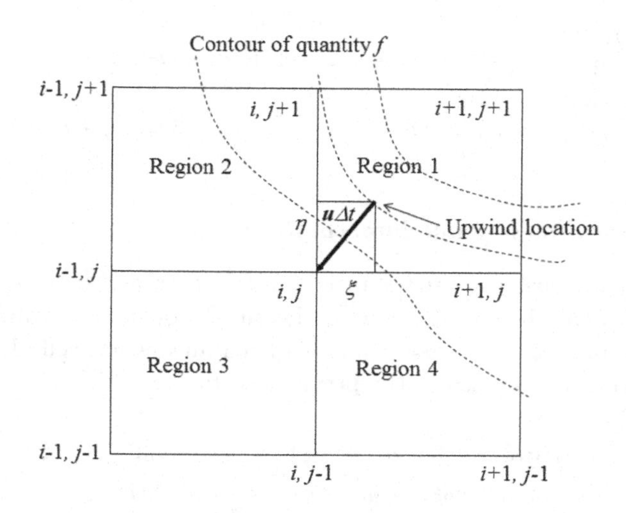

Fig. 5.4. Upwind cells to perform the two dimensional interpolation.

with grid points (i,j), $(i+1,j)$, $(i,j+1)$, and $(i+1,j+1)$ (see region 1 in Fig. 5.4). Therefore, the above constants describe the distribution form of f in this case.

For the case where $u_{i,j} < 0$ and $v_{i,j} > 0$, the second quadrant cell with grids (i,j), $(i-1,j)$, $(i,j+1)$, and $(i-1,j+1)$ (region 2 in Fig. 5.4) is selected for the interpolation. The above local constants a_1–a_{10} are redefined by replacing all the grid indices $i+1$ with $i-1$ and also Δx with $-\Delta x$.

In the same way, in the third quadrant where $u_{i,j} > 0$ and $v_{i,j} > 0$ (region 3 in Fig. 5.4), the indices of the constants are redefined as: $i+1 \to i-1$, $j+1 \to j-1$, and $\Delta x \to -\Delta x$, $\Delta y \to -\Delta y$.

In the fourth quadrant, for $u_{i,j} < 0$ and $v_{i,j} > 0$ (region 4 in Fig. 5.4), $j+1 \to j-1$ and $\Delta y \to -\Delta y$.

The physical value f^{n+1} and its derivatives ∇f^{n+1} at grid point (i,j) at the $n+1$ time step are updated from the distribution forms of both f^n and ∇f^n constrained at the grids of the upwind cell:

$$f_{i,j}^{n+1} = \{(a_1\xi + a_2\eta + a_3)\,\xi + a_4\eta + a_5\}\,\xi + \{(a_6\eta + a_7\xi + a_8)\,\eta + a_9\}\,\eta + a_{10} \tag{5.20}$$

$$\frac{\partial f_{i,j}^{n+1}}{\partial x} = (3a_1\xi + 2a_2\eta + 2a_3)\,\xi + (a_4 + a_7\eta)\,\eta + a_5 \tag{5.21}$$

$$\frac{\partial f_{i,j}^{n+1}}{\partial y} = (a_2\xi + 2a_7\eta + a_4)\,\xi + (2a_8 + 3a_6\eta)\,\eta + a_9 \tag{5.22}$$

5.1.2. *Three-dimensional interpolation*

In the three-dimensional Cartesian coordinate system $(x_{i,j,k}, y_{i,j,k}, z_{i,j,k})$ (see Fig. 5.5), the three-dimensional polynomials for interpolating $f(\boldsymbol{\xi})$ at a local coordinate $\boldsymbol{\xi} = \boldsymbol{x} - \boldsymbol{x}_i = (\xi, \eta, \zeta)$ from an upwind cell also follows the same manner discussed in the previous section:

$$\begin{aligned} f(\xi,\eta,\zeta) = {} & \{(a_1\xi + a_2\eta + a_3\zeta + a_4)\,\xi + a_5\eta + a_6\zeta + a_7\}\,\xi \\ & + \{(a_8\eta + a_9\xi + a_{10}\zeta + a_{11})\,\eta + a_{12}\zeta + a_{13}\}\,\eta \\ & + \{(a_{14}\zeta + a_{15}\xi + a_{16}\eta + a_{17})\,\zeta + a_{18}\}\,\zeta + a_{19}\xi\eta\zeta + a_{20} \end{aligned} \tag{5.23}$$

The first derivatives $\nabla f(\boldsymbol{\xi})$ are thus given as

$$\begin{aligned} \frac{\partial f(\xi,\eta,\zeta)}{\partial x} = {} & (3a_1\xi + 2a_2\eta + 2a_3\zeta + 2a_4)\,\xi \\ & + (a_5 + a_9\eta + a_{19}\zeta)\,\eta + (a_6 + a_{15}\zeta)\,\zeta + a_7 \end{aligned} \tag{5.24}$$

$$\begin{aligned} \frac{\partial f(\xi,\eta,\zeta)}{\partial y} = {} & (3a_8\eta + 2a_9\xi + 2a_{10}\zeta + 2a_{11})\,\eta \\ & + (a_2\xi + a_5 + a_{19}\zeta)\,\xi + (a_{12} + a_{16}\zeta)\,\zeta + a_{13} \end{aligned} \tag{5.25}$$

$$\begin{aligned} \frac{\partial f(\xi,\eta,\zeta)}{\partial z} = {} & (3a_{14}\xi + 2a_{15}\xi + 2a_{16}\eta + 2a_{17})\,\zeta \\ & + (a_3\xi + a_6 + a_19\eta)\,\xi + (a_10\eta + a_{12})\,\eta + a_{18} \end{aligned} \tag{5.26}$$

For the case with advection velocity $u_{i,j,k} < 0$, $v_{i,j,k} < 0$ and $w_{i,j,k} < 0$, i.e., the horizontal upwind displacement $\xi = x - x_i = -u_{i,j,k}\Delta t > 0$, the lateral displacement $\eta = y - y_i = -v_{i,j,k}\Delta t > 0$ and the vertical displacement $\zeta = z - z_i = -w_{i,j,k}\Delta t > 0$ (see Fig. 5.5). The interpolation equations Eqs. (5.23)–(5.26) for the given constraints at the grid points of the upwind hexahedron cell are

$$\begin{aligned} f(l\Delta x, m\Delta y, n\Delta z) &= f_{i+l,j+m,k+n} \quad l,m,n = 0,1 \\ \nabla f(l\Delta x, m\Delta y, n\Delta z) &= \nabla f_{i+l,j+m,k+n} \; l,m,n = 0,1 \end{aligned} \tag{5.27}$$

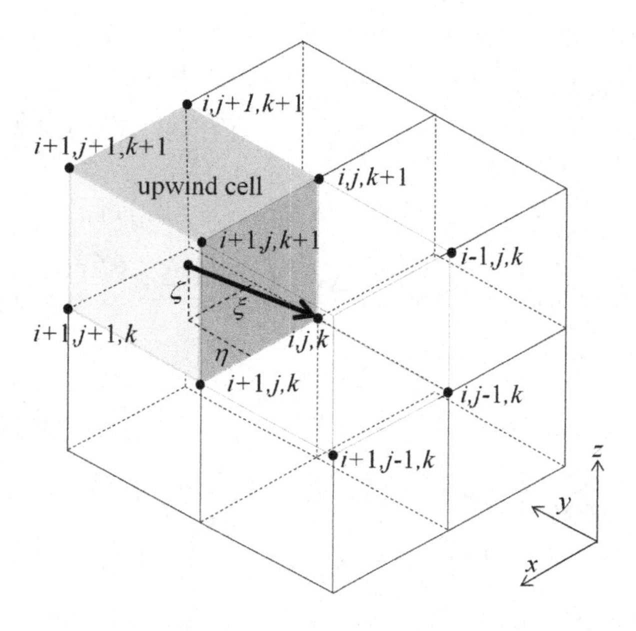

Fig. 5.5. Upwind cells to perform the three dimensional interpolation.

Substituting the above constraints into Eqs. (5.23)–(5.26), twenty local constants a_1 to a_{20} are determined as

$$a_1 = \frac{1}{\Delta x^3}\left\{2\left(f_{i,j,k} - f_{i+1,j}\right) + \left(\frac{\partial f_{i,j,k}}{\partial x} + \frac{\partial f_{i+1,j,k}}{\partial x}\right)\Delta x\right\}$$

$$a_2 = \frac{1}{\Delta x^2 \Delta y}\left\{f_{i,j,k} - f_{i,j+1,k} - f_{i+1,j,k} + f_{i+1,j+1,k} + \left(\frac{\partial f_{i,j,k}}{\partial x} - \frac{\partial f_{i,j+1,k}}{\partial x}\right)\Delta x\right\}$$

$$a_3 = \frac{1}{\Delta x^2 \Delta z}\left\{f_{i,j,k} - f_{i,j,k+1} - f_{i+1,j,k} + f_{i+1,j,k+1} + \left(\frac{\partial f_{i,j,k}}{\partial x} - \frac{\partial f_{i,j,k+1}}{\partial x}\right)\Delta x\right\}$$

$$a_4 = \frac{1}{\Delta x^2}\left\{3\left(f_{i+1,j,k} - f_{i,j,k}\right) - \left(2\frac{\partial f_{i,j,k}}{\partial x} + \frac{\partial f_{i+1,j,k}}{\partial x}\right)\Delta x\right\}$$

$$
\begin{aligned}
a_5 &= \frac{1}{\Delta x \Delta y}\left\{-f_{i,j,k} + f_{i,j+1,k} + f_{i+1,j,k} - f_{i+1,j+1,k}\right. \\
&\quad \left. + \left(-\frac{\partial f_{i,j,k}}{\partial x} + \frac{\partial f_{i,j+1,k}}{\partial x}\right)\Delta x + \left(-\frac{\partial f_{i,j,k}}{\partial y} + \frac{\partial f_{i+1,j,k}}{\partial y}\right)\Delta y\right\} \\
a_6 &= \frac{1}{\Delta x \Delta z}\left\{-f_{i,j,k} + f_{i,j,k+1} + f_{i+1,j,k} - f_{i+1,j,k+1}\right. \\
&\quad \left. + \left(-\frac{\partial f_{i,j,k}}{\partial x} + \frac{\partial f_{i,j,k+1}}{\partial x}\right)\Delta x + \left(-\frac{\partial f_{i,j,k}}{\partial z} + \frac{\partial f_{i+1,j,k}}{\partial z}\right)\Delta z\right\} \\
a_7 &= \frac{\partial f_{i,j,k}}{\partial x} \\
a_8 &= \frac{1}{\Delta y^3}\left\{2\left(f_{i,j,k} - f_{i,j+1,k}\right) + \left(\frac{\partial f_{i,j,k}}{\partial y} + \frac{\partial f_{i,j+1,k}}{\partial y}\right)\Delta y\right\} \\
a_9 &= \frac{1}{\Delta x \Delta y^2}\left\{f_{i,j,k} - f_{i,j+1,k} - f_{i+1,j,k} + f_{i+1,j+1,k}\right. \\
&\qquad\qquad \left. + \left(\frac{\partial f_{i,j,k}}{\partial y} - \frac{\partial f_{i+1,j,k}}{\partial y}\right)\Delta y\right\} \\
a_{10} &= \frac{1}{\Delta z \Delta y^2}\left\{f_{i,j,k} - f_{i,j,k+1} - f_{i,j+1,k} + f_{i,j+1,k+1}\right. \\
&\qquad\qquad \left. + \left(\frac{\partial f_{i,j,k}}{\partial y} - \frac{\partial f_{i,j,k+1}}{\partial y}\right)\Delta y\right\} \\
a_{11} &= \frac{1}{\Delta y^2}\left\{3\left(f_{i,j+1,k} - f_{i,j,k}\right) - \left(2\frac{\partial f_{i,j,k}}{\partial y} + \frac{\partial f_{i,j+1,k}}{\partial y}\right)\Delta y\right\} \\
a_{12} &= \frac{1}{\Delta y \Delta z}\left\{-f_{i,j,k} + f_{i,j,k+1} + f_{i,j+1,k} - f_{i,j+1,k+1}\right. \\
&\quad \left. + \left(-\frac{\partial f_{i,j,k}}{\partial y} + \frac{\partial f_{i,j,k+1}}{\partial y}\right)\Delta x + \left(-\frac{\partial f_{i,j,k}}{\partial z} + \frac{\partial f_{i,j+1,k}}{\partial z}\right)\Delta z\right\} \\
a_{13} &= \frac{\partial f_{i,j,k}}{\partial y} \\
a_{14} &= \frac{1}{\Delta z^3}\left\{2\left(f_{i,j,k} - f_{i,j,k+1}\right) + \left(\frac{\partial f_{i,j,k}}{\partial z} + \frac{\partial f_{i,j,k+1}}{\partial z}\right)\Delta z\right\} \\
a_{15} &= \frac{1}{\Delta x \Delta z^2}\left\{f_{i,j,k} - f_{i,j,k+1} - f_{i+1,j,k} + f_{i+1,j,k+1}\right. \\
&\qquad\qquad \left. + \left(\frac{\partial f_{i,j,k}}{\partial z} - \frac{\partial f_{i+1,j,k}}{\partial z}\right)\Delta z\right\} \\
a_{16} &= \frac{1}{\Delta y \Delta z^2}\left\{f_{i,j,k} - f_{i,j,k+1} - f_{i,j+1,k} + f_{i,j+1,k+1}\right. \\
&\qquad\qquad \left. + \left(\frac{\partial f_{i,j,k}}{\partial z} - \frac{\partial f_{i,j+1,k}}{\partial z}\right)\Delta z\right\}
\end{aligned}
$$

$$a_{17} = \frac{1}{\Delta z^2}\left\{3\left(f_{i,j,k+1} - f_{i,j,k}\right) - \left(2\frac{\partial f_{i,j,k}}{\partial z} + \frac{\partial f_{i,j,k+1}}{\partial z}\right)\Delta z\right\}$$

$$a_{18} = \frac{\partial f_{i,j,k}}{\partial z}$$

$$a_{19} = \frac{1}{\Delta x \Delta y \Delta z}\Big\{-f_{i,j,k} + f_{i,j,k+1} + f_{i,j+1,k} - f_{i,j+1,k+1}$$
$$+ f_{i+1,j,k} - f_{i+1,j,k+1} - f_{i+1,j+1,k} + f_{i+1,j+1,k+1}\Big\}$$

$$a_{20} = \partial f_{i,j,k}$$

As with a two dimensional CIP, these constraints must be imposed at the nodes of the upwind cell depending on the advection velocity; that is, the grid indices and grid widths in the above constants are replaced as follows.

$$\begin{array}{lll} i+1 \to i-1 \;\&\; \Delta x \to -\Delta x, & \text{if} & u_{i,j,k} > 0 \\ j+1 \to j-1 \;\&\; \Delta y \to -\Delta y, & \text{if} & v_{i,j,k} > 0 \\ k+1 \to k-1 \;\&\; \Delta z \to -\Delta z, & \text{if} & w_{i,j,k} > 0 \end{array} \tag{5.28}$$

Assuming a local steady advection process for a short duration Δt, $f_{i,j,k}^{n+1}$ and $\boldsymbol{\nabla} f_{i,j,k}^{n+1}$ at time $t = (n+1)\Delta t$ are updated to be identical with the interpolated values at the local upwind coordinate $\boldsymbol{\xi} = (-u_{i,j,k}\Delta t, -v_{i,j,k}\Delta t, -w_{i,j,k}\Delta t)$ at time $t = n\Delta t$ by Eq. (5.23):

$$\begin{aligned} f_{i,j,k}^{n+1} &= \{(a_1\xi + a_2\eta + a_3\zeta + a_4)\,\xi + a_5\eta + a_6\zeta + a_7\}\,\xi \\ &+ \{(a_8\eta + a_9\xi + a_{10}\zeta + a_{11})\,\eta + a_{12}\zeta + a_{13}\}\,\eta \\ &+ \{(a_{14}\zeta + a_{15}\xi + a_{16}\eta + a_{17})\,\zeta + a_{18}\}\,\zeta + a_{19}\xi\eta\zeta + a_{20} \end{aligned} \tag{5.29}$$

and the first derivatives are also computed as

$$\begin{aligned} \frac{\partial f_{i,j,k}^{n+1}}{\partial x} &= (3a_1\xi + 2a_2\eta + 2a_3\zeta + 2a_4)\,\xi \\ &+ (a_5 + a_9\eta + a_{19}\zeta)\,\eta + (a_6 + a_{15}\zeta)\,\zeta + a_7 \end{aligned} \tag{5.30}$$

$$\begin{aligned} \frac{\partial f_{i,j,k}^{n+1}}{\partial y} &= (3a_8\eta + 2a_9\xi + 2a_{10}\zeta + 2a_{11})\,\eta \\ &+ (a_2\xi + a_5 + a_{19}\zeta)\,\xi + (a_{12} + a_{16}\zeta)\,\zeta + a_{13} \end{aligned} \tag{5.31}$$

$$\begin{aligned} \frac{\partial f_{i,j,k}^{n+1}}{\partial z} &= (3a_{14}\xi + 2a_{15}\xi + 2a_{16}\eta + 2a_{17})\,\zeta \\ &+ (a_3\xi + a_6 + a_{19}\eta)\,\xi + (a_{10}\eta + a_{12})\,\eta + a_{18} \end{aligned} \tag{5.32}$$

5.2. Computational Procedure for the Momentum Equation

Fluid flow is governed by the Navier–Stokes equation, which is composed of advection (inertial), pressure and viscous terms, as explained in §1.1.3. While the CIP method is used to solve only the advection term in this equation in a semi-Lagrangian way, the other terms should be separately computed in an Eulerian numerical approach. A standard procedure applying the CIP method for computations of the Navier–Stokes equation is explained in this section.

Kim and Moin[8] presented a fractional-step method based on the decomposition of multiple operators in the momentum equation for time-dependent incompressible flows. While there are many variations of decomposition and time-advancing schemes in this method (see details in Ferziger and Perić, 1999), CIP is commonly used at the advection step during the fractional-step procedures.

The Navier–Stokes equation for an incompressible fluid

$$\frac{\partial \boldsymbol{u}}{\partial t} = -\left(\boldsymbol{\nabla}\boldsymbol{u}\right)\boldsymbol{u} + \nu\boldsymbol{\nabla}^2\boldsymbol{u} - \frac{1}{\rho}\boldsymbol{\nabla}p \tag{5.33}$$

is expressed in a discrete symbolic form by

$$\frac{\boldsymbol{u}^{n+1} - \boldsymbol{u}^n}{\Delta t} = \boldsymbol{A} + \boldsymbol{D} + \boldsymbol{P} \tag{5.34}$$

where $\boldsymbol{A}$, $\boldsymbol{D}$, and $\boldsymbol{P}$ represent the advection, diffusion, and pressure operators, respectively. The computing procedure for this equation may be decomposed into three steps:

$$\frac{\boldsymbol{u}^{*} - \boldsymbol{u}^n}{\Delta t} = \boldsymbol{A} \tag{5.35}$$

$$\frac{\boldsymbol{u}^{**} - \boldsymbol{u}^{*}}{\Delta t} = \boldsymbol{D} \tag{5.36}$$

$$\frac{\boldsymbol{u}^{n+1} - \boldsymbol{u}^{**}}{\Delta t} = \boldsymbol{P} \tag{5.37}$$

where the superscripts * and ** indicate an intermediate step. $\boldsymbol{P}$ is required to satisfy the divergence-free condition; that is, the pressure should be a solution of the Poisson equation, which is a divergence of Eq. (5.37) imposed on the continuity condition $\boldsymbol{\nabla} \cdot \boldsymbol{u}^{n+1} = 0$

$$\left(\frac{\delta^2}{\delta x^2} + \frac{\delta^2}{\delta y^2} + \frac{\delta^2}{\delta z^2}\right)p = \boldsymbol{\nabla} \cdot \boldsymbol{u}^{**} \tag{5.38}$$

where $\delta/\delta x$ is the discrete finite difference operator. Specific procedures to solve this pressure equation are explained in §3.3.

The CIP method is incorporated into the first step to solve the advection equation. Although the order to solving Eqs. (5.35), (5.36), and (5.37) is interchangeable, the CIP computation at the first step gives more conserved results. A variety of computational approaches are possible for Eqs. (5.36) and (5.37). For instance, when a second order predictor-corrector method based on the trapezoid rule is employed, the procedure is as follows (see also §3.1):

$$\boldsymbol{u}^* = \mathrm{CIP1}\left(\boldsymbol{u}^n\right) \tag{5.39}$$

$$\boldsymbol{\nabla}\boldsymbol{u}^* = \mathrm{CIP2}\left(\boldsymbol{u}^n\right) \tag{5.40}$$

$$\frac{\boldsymbol{u}^{**} - \boldsymbol{u}^*}{\Delta t} = \boldsymbol{D}^* + \boldsymbol{P}^* \tag{5.41}$$

$$\frac{\boldsymbol{u}^{n+1} - \boldsymbol{u}^{**}}{\Delta t} = \frac{1}{2}\left(\boldsymbol{D}^* + \boldsymbol{P}^* + \boldsymbol{D}^{**} + \boldsymbol{P}^{**}\right) \tag{5.42}$$

where CIP1 and CIP2 represent the advection operators for Eq. (5.29) and Eqs. (5.30)–(5.32), respectively (or Eq. (5.20) and Eqs. (5.22) in a two-dimensional CIP). Differentiating Eqs. (5.41) and (5.42) with respect to x and discretizing the right hand side by the central finite difference, we have

$$\begin{aligned}\frac{\partial \boldsymbol{u}^{**}_{i,j,k}}{\partial x} &= \frac{\partial \boldsymbol{u}^{*}_{i,j,k}}{\partial x} + \Delta t\left\{(\mathrm{right})^*_{i+1,j,k} - (\mathrm{right})^*_{i-1,j,k}\right\}/2\Delta x \\ &= \frac{\partial \boldsymbol{u}^{*}_{i,j,k}}{\partial x} + \left(\boldsymbol{u}^{**}_{i+1,j,k} - \boldsymbol{u}^{**}_{i-1,j,k} - \boldsymbol{u}^{*}_{i+1,j,k} + \boldsymbol{u}^{*}_{i-1,j,k}\right)/2\Delta x\end{aligned} \tag{5.43}$$

$$\begin{aligned}\frac{\partial \boldsymbol{u}^{n+1}_{i,j,k}}{\partial x} &= \frac{\partial \boldsymbol{u}^{**}_{i,j,k}}{\partial x} + \Delta t\left\{(\mathrm{right})^{**}_{i+1,j,k} - (\mathrm{right})^{**}_{i-1,j,k}\right\}/2\Delta x \\ &= \frac{\partial \boldsymbol{u}^{**}_{i,j,k}}{\partial x} + \left(\boldsymbol{u}^{n+1}_{i+1,j,k} - \boldsymbol{u}^{n+1}_{i-1,j,k} - \boldsymbol{u}^{**}_{i+1,j,k} + \boldsymbol{u}^{**}_{i-1,j,k}\right)/2\Delta x\end{aligned} \tag{5.44}$$

Substituting Eq. (5.43) into Eq. (5.44),

$$\frac{\partial \boldsymbol{u}^{n+1}_{i,j,k}}{\partial x} = \frac{\partial \boldsymbol{u}^{*}_{i,j,k}}{\partial x} + \left(\boldsymbol{u}^{n+1}_{i+1,j,k} - \boldsymbol{u}^{n+1}_{i-1,j,k} - \boldsymbol{u}^{*}_{i+1,j,k} + \boldsymbol{u}^{*}_{i-1,j,k}\right)/2\Delta x \tag{5.45}$$

The other spatial derivatives are also obtained in the same way. The velocity gradients in the Eulerian stage are updated by the above equation using both the velocity and its derivatives that are explicitly updated by the CIP in a Lagrangian stage in Eqs. (5.39) and (5.40).

5.3. Extension of the CIP

While the standard CIP is interpreted in the previous section, there have been many other versions of CIP methods: CIP using higher order polynomials[1,2] , CIP using an integrated form of polynomials (CIPCLS method that will be described later),[15,16,25,37] CIP with a unified solution technique for both liquid and gas flows (CCUP method),[34] and so on. In this section, CCUP and CIPCLS, which have been applied to wave dynamics, are outlined.

5.3.1. *CIP Combined, Unified Procedure (CCUP)*

Several unified solution techniques for compressible and incompressible two-phase fluids have been proposed (e.g. Hirt and Nichols[5]). Yabe and Wang[34] also developed a CIP based technique for computing gas-liquid two-phase flows.

Mass conservation for compressible flow is given by

$$\frac{\partial \rho}{\partial t} + \boldsymbol{\nabla} \cdot (\rho \boldsymbol{u}) = 0 \tag{5.46}$$

where ρ is the fluid density. This is also written as

$$\frac{d\rho}{dt} + \rho \boldsymbol{\nabla} \cdot \boldsymbol{u} = 0 \tag{5.47}$$

Assuming an adiabatic process of isentropic flow, Eq. (5.47) may be rewritten as

$$\frac{1}{dp/d\rho} \frac{dp}{dt} + \rho \boldsymbol{\nabla} \cdot \boldsymbol{u} = 0 \tag{5.48}$$

where $dp/d\rho = c^2 = \gamma p/\rho$, c is the speed of sound, and γ is the adiabatic index. Using the fractional-step method for the discretized form of Eq. (5.48), we have

$$p^* = \mathrm{CIP1}\,(p^n) \tag{5.49}$$

$$\begin{aligned} p^{n+1} &= p^* - c^2 \rho \boldsymbol{\nabla} \cdot \boldsymbol{u}^{n+1} \Delta t \\ &= p^* - \gamma p^* \boldsymbol{\nabla} \cdot \boldsymbol{u}^{n+1} \Delta t \end{aligned} \tag{5.50}$$

The advection of pressure is computed by using the CIP method in Eq. (5.49), which may be omitted if $u/c \ll 1$. The fractional-step method is also applied to the momentum equation for compressible flow in the same manner as outlined in §5.2.

$$\boldsymbol{u}^* = \mathrm{CIP1}\left(\boldsymbol{u}^n\right) \tag{5.51}$$

$$\frac{\boldsymbol{u}^{**} - \boldsymbol{u}^*}{\Delta t} = \boldsymbol{D} \tag{5.52}$$

$$\frac{\boldsymbol{u}^{n+1} - \boldsymbol{u}^{**}}{\Delta t} = \boldsymbol{P} = -\frac{\boldsymbol{\nabla} p^{n+1}}{\rho^*} \tag{5.53}$$

The Poisson pressure equation obtained by taking the divergence of Eq. (5.53) is expressed as

$$\left(\frac{\delta}{\delta x^2} + \frac{\delta}{\delta y^2} + \frac{\delta}{\delta z^2}\right)\left(\frac{p^{n+1}}{\rho^*}\right) = -\frac{1}{\Delta t}\left(\boldsymbol{\nabla}\cdot\boldsymbol{u}^{n+1} - \boldsymbol{\nabla}\cdot\boldsymbol{u}^{**}\right) \tag{5.54}$$

Substituting Eq. (5.50) into Eq. (5.54), we have

$$\begin{aligned}\left(\frac{\delta^2}{\delta x^2} + \frac{\delta^2}{\delta y^2} + \frac{\delta^2}{\delta z^2}\right)\left(\frac{p^{n+1}}{\rho^*}\right) &= \frac{p^{n+1} - p^*}{\Delta t^2 \rho^* c^2} + \frac{1}{\Delta t}\boldsymbol{\nabla}\cdot\boldsymbol{u}^{**} \\ &= \frac{p^{n+1} - p^*}{\Delta t^2 \gamma p^*} + \frac{1}{\Delta t}\boldsymbol{\nabla}\cdot\boldsymbol{u}^{**}\end{aligned} \tag{5.55}$$

For the case with incompressible flow, Eq. (5.55) is found to be equivalent to Poisson equation (5.38) because $c \to \infty$; that is, iterative computations of Eq. (5.55) determine the simultaneous pressures in both the compressible and incompressible fluids at equilibrium.

To calculate the density, the discrete form of Eq. (5.47) is decomposed into

$$\rho_i^* = \mathrm{CIP1}\left(\rho^n\right) \tag{5.56}$$

$$\frac{\rho^{n+1} - \rho^*}{\Delta t} = -\rho^*\boldsymbol{\nabla}\cdot\boldsymbol{u}^{n+1} \tag{5.57}$$

Eq. (5.57) is modified by Eq. (5.50):

$$\rho^{n+1} - \rho^* = \frac{\rho^*}{\gamma p^*}\left(p^{n+1} - p^*\right) \tag{5.58}$$

The fluid density can be explicitly updated using the pressure obtained from Poisson equation (5.55).

The coupled velocity, pressure, and density values in gas-liquid flows are computed by the entire set of the above discretized equations through a combination of CIP and fractional-step methods.

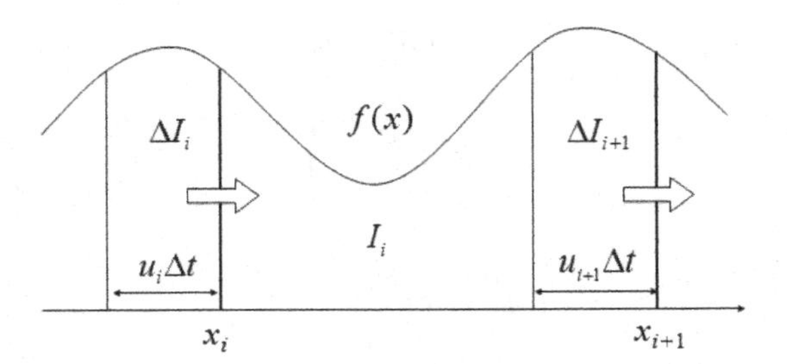

Fig. 5.6. Definition of an integrated quantity I and ΔI in a computing cell.

5.3.2. *CIP-Conservative semi-Lagrangian Scheme (CLS)*

The standard CIP method defines a spatial distribution f and its derivatives ∇f within a upwind cell to update them. In CIPCLS, an integrated form of the quantity is introduced:

$$I_i = \int_{x_i}^{x_{i+1}} f(x)\, dx \tag{5.59}$$

The integrated quantity I is also expressed in finite difference form as

$$I_i^{n+1} = I_i^n + \Delta I_{i+1} - \Delta I_i \tag{5.60}$$

where ΔI_i is the flux across x_i during Δt and is defined as

$$\Delta I_i = \int_{x_i}^{x_i+\xi} f(x)\, dx \tag{5.61}$$

where $\xi = -u_i \Delta t$ (see Fig. 5.6).

It is readily found that the total quantity integrated over the domain $\sum_i I_i$ is conserved because the sum of ΔI_i over all the cells is always zero. The advection equations for the integrated quantity $D = \int f dx$ and its derivative $D' = f$ are considered in this method.

$$\frac{\partial D}{\partial t} + u\frac{\partial D}{\partial x} = 0 \tag{5.62}$$

Differentiating the above equation with respect to x,

$$\frac{\partial D'}{\partial t} + \frac{\partial u D'}{\partial x} = 0 \tag{5.63}$$

where the integrated quantity D_i is defined as

$$D_i(x) = \int_{x_i}^{x} f(\xi)\, d\xi \tag{5.64}$$

D is interpolated in the cell with cubic polynomials:

$$D_i(x) = a\xi^3 + b\xi^2 + f_i^n \xi \tag{5.65}$$

where the upwind coordinate $\xi = x - x_i$. Eq. (5.65) is required to satisfy the constraints at the grid cell:

$$D_i(x_i) = 0 \tag{5.66}$$

$$D_i(x_{iup}) = \text{sign}(\xi) I_{icell}^n \tag{5.67}$$

where I_{icell}^n is the quantity integrated from x_i to the upwind grid location x_{iup}, which is obtained from Eq. (5.59); and $icell = i + (\text{sign}(\xi) - 1)/2$.

Differentiating Eq. (5.65), the spatial distribution of $f(x)$ is given by

$$f(x) = \frac{\partial D_i(x)}{\partial x} = 3a\xi^2 + 2b\xi + f_i^n \tag{5.68}$$

The first derivative of D_i is also constrained at the grid:

$$\frac{\partial D_i(x_i)}{\partial x} = f_i \tag{5.69}$$

$$\frac{\partial D_i(x_{iup})}{\partial x} = f_{iup} \tag{5.70}$$

The local constants a and b in Eq. (5.68) under those constraints are determined by

$$a = \frac{f_i^n + f_{iup}^n}{\Delta x^2} - \frac{2\text{sign}(\xi)\rho_{icell}^n}{\Delta x^3} \tag{5.71}$$

$$b = -\frac{2f_i^n + f_{iup}^n}{\Delta x} + \frac{3\text{sign}(\xi)\rho_{icell}^n}{\Delta x^2} \tag{5.72}$$

where the grid width $\Delta x = x_{iup} - x_i$. ΔI_i can be written from Eq. (5.64) as

$$\Delta I_i = \int_{x_i}^{x_i+\xi} f(x)\,dx = D_i(x_i + \xi) = a\xi^3 + b\xi^2 + f_i^n \xi \tag{5.73}$$

Substituting ΔI_i estimated by the above equation into Eq. (5.60), the integrated quantity I_i^{n+1} can be updated.

5.4. Computation of Free-Surfaces and Interfaces

Computational strategies to reproduce free-surface or interface dynamics are crucially important when using a discrete grid system to reasonably approximate discontinuous density values and stresses at surfaces. The computational elements to consider include choosing a grid system and a computational scheme to detect and track surfaces. Although many computational methods to model and approximate surface flows have been proposed (see a review by Scardovelli and Zaleski[21]), solving differential equations at the surface, which always becomes anisotropic and mechanically singular, is still very difficult and challenging.

In this section, fundamental procedures and techniques to compute surface locations and nearby fluid flows are explained.

The location of a surface ζ_i is transported as

$$\frac{D\zeta_i}{Dt} = \frac{\partial \zeta_i}{\partial t} + u_j \frac{\partial \zeta_i}{\partial x_j} = 0 \tag{5.74}$$

Since this nonlinear advection equation does not consider any continuity conditions connected with inner fluid flow, poor computations of this equation often cause two notable problems: unstable pressure oscillations resulting from surface locations that separate from the inner fluid, and numerical mass-loss caused by computed surfaces penetrating the inner fluid region. As will be explained in Chapter 4, with the VOF method, the above equation is not explicitly solved, and the surface location is reconstructed by volume fractions of fluid computed from the mass flux into and out of the cell via a donor-acceptor method. While this method is consistent with the conservation of fluid volume across the cells, significant mass-loss may occur at highly curved surfaces that are poorly reconstructed from the discrete volume fractions. With these types of surfaces, there is no guarantee that the computed surface profile is still consistent with the net mass flux across the cell surface. When this occurs, the surface location within the cell needs to be interpolated with higher order polynomials for accurate reconstruction.[21]

To compute surface flows, there are interdependent procedures to numerically define the surfaces and to model discontinuous surface stresses. One such procedure involves using adaptive grids; grids and cell-faces are computed, with the advantage that the cell-faces coincide with the surface locations where dynamic boundary conditions are imposed. However, there are difficulties generating arbitrary grids that can adapt to the complex shapes of fragmenting or coalescing surfaces that are often observed in a wave breaking event. On the other hand, in a fixed grid system, there is no concern about grid generation, even though the subgrid surface locations require precise definition where discrete surface stresses are imposed within the cell. The numerical procedures involving the CIP method with a fixed grid system are described in this section.

5.4.1. *Definition of surfaces*

A kinematic boundary condition at a surface is expressed as

$$\frac{DF}{Dt} = \frac{\partial F}{\partial t} + u_j \frac{\partial F}{\partial x_j} = 0 \tag{5.75}$$

where the arbitrary function $F(x_i, t)$ defines the surface location. For a free-surface flow, a density function (or color function) $\rho_f(x_i, t)$ may be introduced as the surface function to indicate ρ_1 in a liquid region and ρ_0 on the exterior of the fluid. For an interfacial flow of two immiscible fluids, ρ_f is defined to be ρ_1 in one fluid and ρ_0 in the other. The free-surface or interface is defined to be at the location where $\rho_s = (\rho_1 + \rho_0)/2$. Substituting $F = \rho_f$ into Eq. (5.75)

$$\frac{D\rho_f}{Dt} = \frac{\partial \rho_f}{\partial t} + u_j \frac{\partial \rho_f}{\partial x_j} = 0 \tag{5.76}$$

The density function method determines that the surface location will be where $\rho_f = \rho_s$, as determined from suitable interpolation between adjacent discrete grids of the computed density functions. A distinct advantage of the CIP method is the ability to compute the advection of ρ_f with high gradients near the surfaces while preserving the subgrid profile within the surface cell. While this simple method can be widely applied to calculate complex surface deformation,[28,29] numerical diffusion of the density function may occur, effectively smearing the surface over computations of long duration.

A tangential transform for the density function, proposed by Yabe and Xiao,[36] can be applied to avoid excessive diffusion and maintain sharp surfaces. In this technique, the following surface function is used for Eq. (5.75) instead of $F = \rho_f$.

$$F = \tan(\pi(\rho_f - 0.5)) \tag{5.77}$$

Osher and Sethian[18] introduced a signed distance function, or a so-called level set function, as a surface function in Eq. (5.75). This level set method has been widely applied to free-surface and interfacial flows in various fields.[19,22,31] The level set function d indicates positive and negative distances from the surface to the interior and exterior of the fluid region, respectively. The surface is thus defined to be at the location $d = 0$. Unlike the density function, which has a high gradient across the surface similar to a Heaviside function, the level set function is smoothly distributed across the domain except for at a singular region where the sign of ∇d reverses. The advection equation used in the level set method is

$$\frac{Dd}{Dt} = \frac{\partial d}{\partial t} + u_j \frac{\partial d}{\partial x_j} = 0 \tag{5.78}$$

The level set function explicitly provides geometric characteristics of the surface; that is, since ∇d is perpendicular to the iso-surface of $d = 0$ (surface), the unit outward normal vector $\boldsymbol{n}$ and surface curvature κ can be estimated with

$$\boldsymbol{n} = -\frac{\nabla d}{|\nabla d|} \tag{5.79}$$

$$\kappa = \nabla \cdot \boldsymbol{n} \tag{5.80}$$

Numerical methods that use scalar functions to identify surfaces during advection computations are generally called front capturing methods. The CIP method is one type of an advantageous scheme to update the surface function in a front capturing approach.

Another explicit type of approach is the front tracking method.[26] With this method, the computational grids adapt to the surfaces and the locations are directly computed from Eq. (5.74) at each time step; the surface grids are explicitly displaced according to the advection of the surface. The front tracking method has been successfully applied to gas-liquid two phase bubble flows (e.g. Esmaeeli and Tryggvason[4]).

5.4.2. *Mechanical balance at surfaces*

Numerical approximations of momentum and fluid stress at free-surfaces and interfaces are described in this section.

Consider an interfacial flow of two immiscible incompressible fluids labeled 1 and 2. The momentum equation for the kth fluid ($k = 1, 2$) is written as

$$\frac{\partial u_i^{(k)}}{\partial t} + u_j^{(k)} \frac{\partial u_i^{(k)}}{\partial x_j} = -\frac{1}{\rho^{(k)}} \frac{\partial p^{(k)}}{\partial x_i} + \frac{\mu^{(k)}}{\rho^{(k)}} \frac{\partial^2 u_i^{(k)}}{\partial x_j \partial x_j} + g_i, \tag{5.81}$$

where μ and g_i are the viscosity coefficient and gravity vector, respectively. Assuming the surface tension σ is uniform along the interface, the momentum conservation across the interface leads to the normal and tangential dynamic boundary conditions:

$$p^{(1)} - p^{(2)} = n_i \tau_{ij}^{(1)} n_j - n_i \tau_{ij}^{(2)} n_j + \sigma\kappa \tag{5.82}$$

$$n_i \tau_{ij}^{(1)} t_j - n_i \tau_{ij}^{(2)} t_j = 0 \tag{5.83}$$

where κ is the curvature, and the viscous stress tensor τ_{ij} is given as

$$\tau_{ij}^{(k)} = \mu^{(k)} \left(\frac{\partial u_i^{(k)}}{\partial x_j} + \frac{\partial u_j^{(k)}}{\partial x_i} \right) \tag{5.84}$$

For the free-surface flow of fluid 1, fluid 2 is no longer taken into account. The above jump conditions are reduced to

$$p^{(1)} = p_0 + n_i \tau_{ij}^{(1)} n_j + \sigma\kappa \tag{5.85}$$

$$n_i \tau_{ij}^{(1)} t_j = 0 \tag{5.86}$$

where p_0 is atmospheric pressure. Normal jump conditions in Eqs. (5.82) and (5.85) define the mechanical balance of pressure, normal viscous stress, and surface tension, which play a substantial role in wind-wave generation, capillary surface motion at a steep wave crest, and wave breaking processes including spray and bubble formation in the surf zone.

The simplest representation of the free-surface, when the second and third terms on the right hand side of Eq. (5.85) are omitted, often occurs with gravity-dominant free-surface flows with negligibly small capillary and viscous effects, such as small amplitude waves and nearshore currents. Then, the boundary condition of Eq. (5.85) is simplified to $p^{(1)} = p_0$, which is explicitly given at the surface location for the Poisson pressure equation (5.38). On the other hand, the tangential boundary condition of Eq. (5.83) determines the wind shear stresses that drive ocean currents, wind waves, and white capping. The zero tangential stress condition of Eq. (5.86) results in vorticity at a curved free-surface,[10] which predominates the vorticity evolution in breaking waves and bores with highly curved free-surfaces.

As explained in the previous section, an advantage of the front tracking approach is the ability to impose these surface boundary conditions because the normal and tangential derivatives of any quantity in both phases can be explicitly computed on the adaptive grids that are aligned with the interface[a]. By contrast, special techniques are needed to impose the jump conditions in the front capturing method used with a fixed Cartesian coordinate system. Also, the velocity and its derivatives both inside and outside of (but adjacent to) the surface are generally needed to solve the advection equation for the surface function in a discrete grid system.

[a]This is achieved with the identical procedure of imposing wall boundary conditions on a collocated grid arrangement with arbitrary control volume shapes, explained in §3.2.1.

To fulfill the dynamic boundary condition of Eq. (5.86) at the surface, practical way to achieve free-surface advection on the fixed grids is to extrapolate the fluid velocity from the interior fluid area to the empty grids outside of the fluid[b]. Popinet and Zaleski[20] presented an extrapolation technique for a two-dimensional liquid-phase velocity calculation using empty cells inside an air bubble, a procedure that implicitly fulfills the zero tangential shear condition at the free-surface Eq. (5.86). With this method, the optimal velocity outside the surface is extrapolated to minimize a penalty term for ensuring the zero tangential stress condition Eq. (5.86). This is a reasonable approach for computing the advection of the free surface using local velocity terms, which approximately satisfies the dynamic boundary condition. The evolution of collapsing bubble shapes has been computed using this method.[20]

Watanabe *et al.*[31] also presented a numerical method which fulfills the free-surface boundary conditions and extrapolates the fluid velocity to empty grids outside the fluid region using a fixed, Cartesian grid system in three-dimensional space. The complex, three-dimensional, vortex structures formed via surface/vortex interaction and induction between vortices have been computed using this proposed technique, implemented within a level-set method for both vertical and oblique droplet impacts in incompressible fluids. This extrapolation technique is outlined below.

The velocity gradients at the free-surface, $\partial u_i^s(\boldsymbol{x}_s)/\partial x_j$, are assumed to be described by the sum of the velocity gradients of the interior surface of the fluid and the correction function f_{ij}

$$\frac{\partial u_i(\boldsymbol{x}_s+\boldsymbol{\xi})}{\partial x_j}=\frac{\partial u_i^s(\boldsymbol{x}_s)}{\partial x_j}+f_{ij}(\boldsymbol{\xi}) \tag{5.87}$$

where $\boldsymbol{\xi}=(\xi,\eta,\zeta)$ is the position vector from the surface location $\boldsymbol{x}_s$. The correction function is chosen to approximate spatial variations of the velocity gradients near the surface. The linear function with local constants $a_{ij}, b_{ij}, c_{ij}, d_{ij}$ is simply used to define the correction function $f_{ij}(\boldsymbol{\xi}) = a_{ij}+b_{ij}\xi+c_{ij}\eta+d_{ij}\zeta$. Substituting Eq. (5.87) into the zero tangential shear condition in Eq. (5.86), we have the approximated conditions for two tangential directions t_i and s_i:

[b]In §3.2.1, the wall boundary condition on staggered grids is satisfied via extrapolation from the interior fluid velocity to the exterior of the domain with a first-order approximation. A higher order extrapolation is required to describe the surface shear and the resulting dynamic deformation of the surface or interface boundary.

$$\sum_{i,j=1,2,3}\left(\left(\frac{\partial u_i(\boldsymbol{x}_s+\boldsymbol{\xi})}{\partial x_j}-f_{ij}(\boldsymbol{\xi})\right)+\left(\frac{\partial u_j(\boldsymbol{x}_s+\boldsymbol{\xi})}{\partial x_i}-f_{ji}(\boldsymbol{\xi})\right)\right)n_j t_i=0 \tag{5.88}$$

$$\sum_{i,j=1,2,3}\left(\left(\frac{\partial u_i(\boldsymbol{x}_s+\boldsymbol{\xi})}{\partial x_j}-f_{ij}(\boldsymbol{\xi})\right)+\left(\frac{\partial u_j(\boldsymbol{x}_s+\boldsymbol{\xi})}{\partial x_i}-f_{ji}(\boldsymbol{\xi})\right)\right)n_j s_i=0. \tag{5.89}$$

The constants $a_{ij}, b_{ij}, c_{ij}, d_{ij}$ used to minimize errors of these equations are determined by using the least squares method. Thus the optimal surface velocity, $u_i^s(\boldsymbol{x}_s)$, that satisfies the condition of Eq. (5.86) is given by the Taylor approximation:

$$\begin{aligned} u_i^s(\boldsymbol{x}_s) &= u_i(\boldsymbol{x}_s+\boldsymbol{\xi})+b_{i1}\xi^2+c_{i2}\eta^2+d_{i3}\zeta^2+(c_{i1}+b_{i2})\xi\eta \\ &\quad +(d_{i1}+b_{i3})\xi\zeta+(d_{i2}+c_{i3})\eta\zeta+\left(a_{i1}-\frac{\partial u_i(\boldsymbol{x}_s+\boldsymbol{\xi})}{\partial x}\right)\xi \\ &\quad +\left(a_{i2}-\frac{\partial u_i(\boldsymbol{x}_s+\boldsymbol{\xi})}{\partial y}\right)\eta+\left(a_{i3}-\frac{\partial u_i(\boldsymbol{x}_s+\boldsymbol{\xi})}{\partial z}\right)\zeta \end{aligned} \tag{5.90}$$

The ghost velocity at the outer empty grid, $u_i(\boldsymbol{x}_s+\boldsymbol{\xi})$, is also determined by extrapolation using the optimal constants and the above surface velocity $u_i^s(\boldsymbol{x}_s)$, which is consistent with the zero tangential shear condition:

$$\begin{aligned} u_i(\boldsymbol{x}_s+\boldsymbol{\xi}) &= u_i^s(\boldsymbol{x}_s)-b_{i1}\xi^2-c_{i2}\eta^2-d_{i3}\zeta^2-(c_{i1}+b_{i2})\xi\eta \\ &\quad -(d_{i1}+b_{i3})\xi\zeta-(d_{i2}+c_{i3})\eta\zeta+\left(\frac{\partial u_i(\boldsymbol{x}_s+\boldsymbol{\xi})}{\partial x}-a_{i1}\right)\xi \\ &\quad +\left(\frac{\partial u_i(\boldsymbol{x}_s+\boldsymbol{\xi})}{\partial y}-a_{i2}\right)\eta+\left(\frac{\partial u_i(\boldsymbol{x}_s+\boldsymbol{\xi})}{\partial z}-a_{i3}\right)\zeta \end{aligned} \tag{5.91}$$

Watanabe *et al.*[31] validated this technique using numerical tests, comparing the results with experimental observations of a vortex ring formed under a droplet that impacts a free surface (see Fig. 5.7).

In the two-phase flow with gas and liquid, the fluid density, viscosity, and spatial derivatives of fluid velocities become discontinuous at the interface. In computations on a fixed grid system, these discontinuities, which are located between discrete grids, have been dealt with using a finite volume method via a shock capturing technique for a Riemann problem (e.g., see Kelecy and Pletcher[7]). However, approximating discontinuous quantities across a sharp interface presents difficulties with numerical stability; non-physical oscillations (Gibbs phenomenon) may develop near the interface.

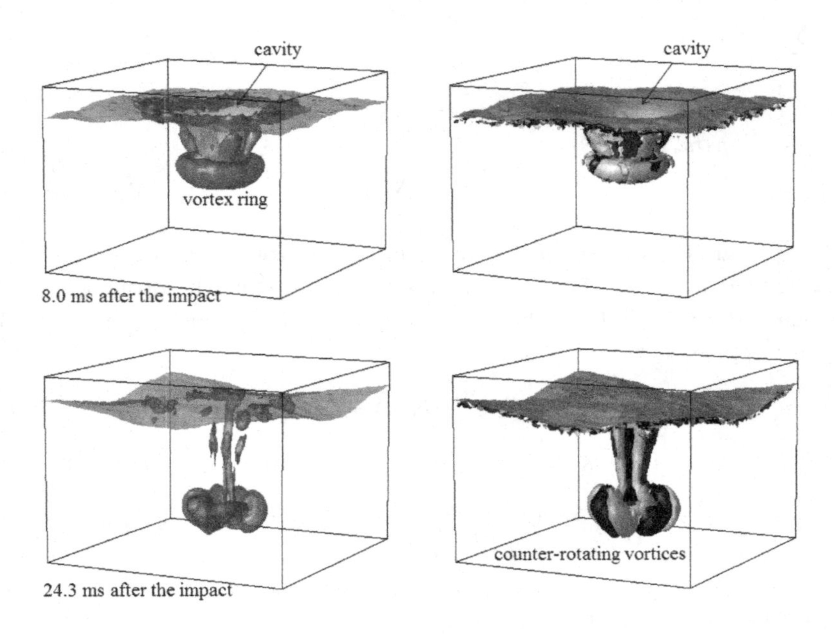

Fig. 5.7. Computed distributions of numerical dye trapped within a vortex ring that forms after impact (left) and iso-surfaces of vorticity showing counter-rotating vertical vortices wrapped by a vortex ring (right);[31] white: dimensionless vorticity of 0.1, black: -0.1.

One possible way to avoid these problems and proceed is to neglect rigorous interfacial dynamics that result from the imposed surface conditions in Eqs. (5.82)–(5.83). And, to suppress oscillations and improve numerical stability, we can smear out the interface over several grids. The fluid density $\rho(x_i)$ and viscosity $\mu(x_i)$ are smeared out within a finite thickness ε covering 3 - 5 grids around the interface:

$$\rho(x_i) = \rho^{(2)} + \left(\rho^{(1)} - \rho^{(2)}\right) H(x_i - \zeta_i) \tag{5.92}$$

$$\mu(x_i) = \mu^{(2)} + \left(\mu^{(1)} - \mu^{(2)}\right) H(x_i - \zeta_i) \tag{5.93}$$

where the numerical Heaviside function smoothly varies with distance from the interface located at ζ_i:

$$H(x_i-\zeta_i)=\begin{cases}0 & \text{if}\quad (x_i-\zeta_i)\,n_i<-\varepsilon\\ \dfrac{1}{2}\left(1+\dfrac{(x_i-\zeta_i)\,n_i}{\varepsilon}+\dfrac{1}{\pi}\sin\left(\dfrac{\pi\,(x_i-\zeta_i)\,n_i}{\varepsilon}\right)\right) & \\ \qquad\qquad \text{if}\quad -\varepsilon\le(x_i-\zeta_i)\,n_i\le\varepsilon & \\ 1 & \text{if}\quad \varepsilon<(x_i-\zeta_i)\,n_i\end{cases}\tag{5.94}$$

Brackbill[3] proposed a volumetric representation for the normal boundary condition of interfaces between inviscid incompressible fluids ($\mu=0$), the so-called continuum surface force (CSF) model. In this model, the surface tension is also smeared out over ε and defined by

$$\sigma\kappa\delta\,(x_i-\zeta_i)\,n_i\tag{5.95}$$

Using these smeared quantities, the momentum equations Eq. (5.81) for multiple phases may thus be rewritten into a single unified equation:

$$\frac{\partial u_i}{\partial t}+u_j\frac{\partial u_i}{\partial x_j}=-\frac{1}{\rho(x_i)}\frac{\partial p}{\partial x_i}+\frac{\mu(x_i)}{\rho(x_i)}\frac{\partial^2 u_i}{\partial x_j\partial x_j}+\sigma\kappa\delta\,(x_i-\zeta_i)\,n_i+g_i\tag{5.96}$$

where the numerical Delta function is defined as

$$\delta(x_i-\zeta_i)=\begin{cases}\dfrac{1}{2\varepsilon}\left(1+\cos\left(\dfrac{\pi|x_i-\zeta_i|}{\varepsilon}\right)\right) & \text{if}\ -\varepsilon\le|x_i-\zeta_i|\le\varepsilon\\ 0 & \text{if}\ \varepsilon<|x_i-\zeta_i|\end{cases}\tag{5.97}$$

This approach suppresses unstable behaviors at the interfaces and unwanted pressure oscillations, and it eliminates the need for interface reconstruction, which are great advantages for practical computations. However, since tangential conditions are not considered in this model, smooth velocity variations over a thick interface (thickness of ε) are computed, which may deviate from the interfacial shear condition of Eq. (5.83) that defines a discontinuous jump in velocity gradients at the interface between fluids with different viscosities. Therefore, this approach may not provide rigorous interfacial dynamics, including surface-vorticity interactions, that are often observed at breaking wave surfaces as shown in Fig. 5.12[c].

[c] While the fluid description of Eqs. (5.92), (5.93) and (5.96) approach the original momentum equation with a mathematically consistent model for surface tension at the limit $\varepsilon\to 0$, there is no physical rationale to validate the smearing procedure with finite ε. One might think there is a very thin region where the density can be approximated by Eq. (5.92), a region in the vicinity of the interface where evaporation and condensation continuously occur in an equilibrium state, called a mass boundary layer or Knudsen

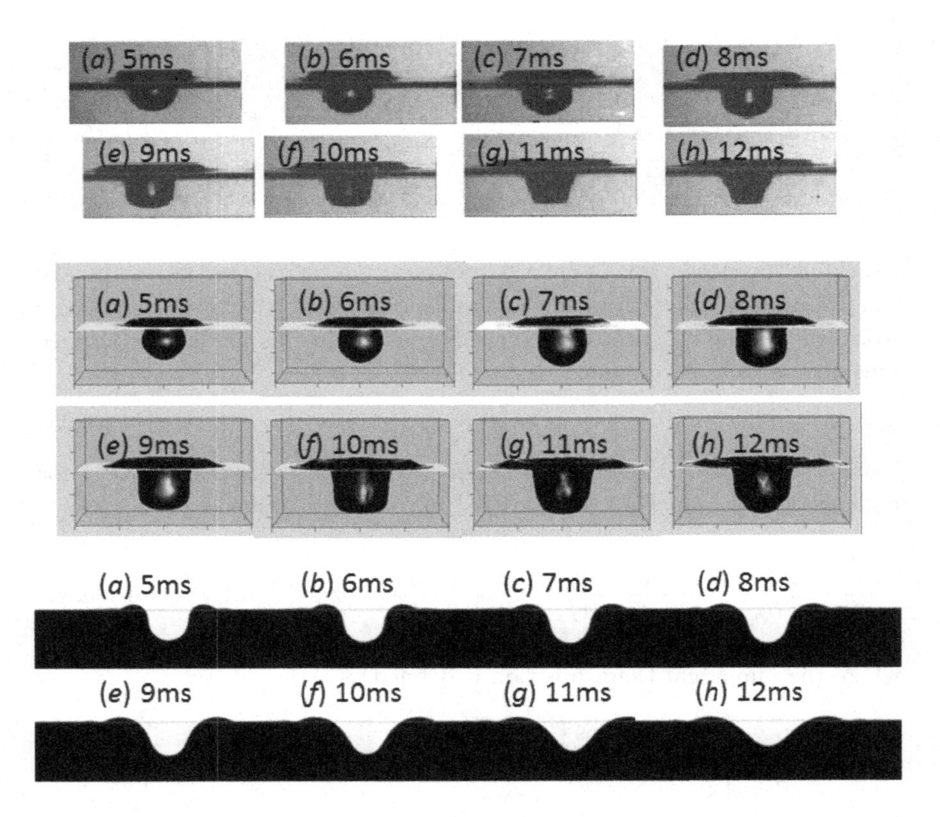

Fig. 5.8. Cavity formations of a droplet after impact with a liquid (drop diameter; 2.12 mmCimpact velocity; 2.13 m/s): experimental observation (top),[11] computed result by the surface model of Watanabe *et al.* (middle),[31] and computed result using the gas-liquid unified momentum equation Eq. (5.96) and the CSF model (bottom).

Figure 5.8 is a typical example showing some distinct differences in the cavity formations as computed by Watanabe's surface boundary model[31] and by the CSF model[3] [d]. While Watanabe's model reasonably describes evolution of the local cavity shape predominated by capillary effects and

---

layer. However, the thickness of such a layer may be $O(1\mu m)$, and therefore much finer grids are required to resolve it, a requirement that is very difficult to achieve in terms of computing cost. Also, evaporation and condensation processes are physically described in terms of thermal dynamics or molecular dynamics and are governed by the Boltzmann equation not Navier–stokes equation.

[d]A open source CFD software package Open FOAM was used for this computation. `http://www.openfoam.com/`

shows agreement with the experimental images of the cavity forms,[11] the CSF model results in a smeared solution; the bottom edges of the computed cavity are much milder than the experimental ones, and the cavity collapses earlier than is observed experimentally. Since emerging research suggests that the surface vorticity is proportional to the surface curvature,[10] reasonable sub-surface vortices, like those shown in Fig. 5.7, would not be expected to form on smeared interfaces with milder deformation, as in this case.

To describe the consistent features of the flow, a suitable surface model should be carefully chosen to compute the flow field.

5.5. Numerical Wave Flume Based on the CIP Method

Precise computations based on the CIP technique have been performed for coastal engineering research since the late 1990s, which have contributed to our understanding of turbulent flows in the coastal surf zone, wave-structure interactions, and sediment transport. Some CIP applications to wave dynamics are reviewed in this section.

In the surf zone, a shoaling wave crest overturns and plunges onto a forward wave face, forming a horizontal roller vortex around an inner surface of the plunging jet, as well as strong turbulence at the plunging locations. As the dynamic splash-up continues in a transition region, successive plunging jets locally intensify turbulence, leading to fully developed turbulence that dominates a bore region. Sea sprays are dispersed and air bubbles are simultaneously entrained into the water through the breaking process. Representing two-phase liquid-air flows and the complex turbulence that results requires comprehensive computations, and several important aspects of the dynamic and turbulent surfaces that evolve in the surf zone have been successfully modeled with Large Eddy Simulations incorporated into the CIP technique.

Shoaling waves with overturning deformation of the free-surface have been precisely reproduced in two-dimensional computations.[29] In three-dimensional computations, wave plunging has exhibited simultaneous spanwise velocity (in the direction of the wave crest) underneath plunging jets, producing rapidly organizing three-dimensional coherent structures.[28] Watanabe *et al.*[30] found that these coherent streamwise and vertical vorticities are initiated from the reorientation of the initial, two-dimensional, spanwise vorticity that forms around an air tube. They are intensified along a strong shear layer that forms between primary horizontal rollers

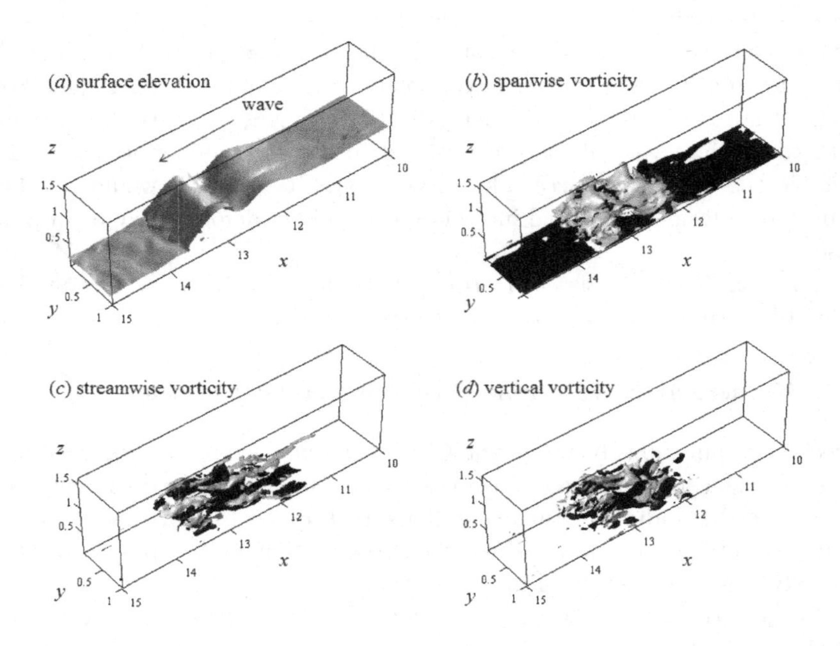

Fig. 5.9. In the surf zone: shape of the free-surface (a), iso-surfaces of spanwise vorticity (b), streamwise vorticity (c), and vertical vorticity (d); white: dimensionless vorticity of 1, black: -1.

and evolve into counter-rotating vortices that stretch obliquely downward (See Fig. 5.9). The resulting coherent rib structure involves longitudinal vortices connected with roller vortices and is developed in repeating splash-up processes (see Fig. 5.10). Watanabe *et al.*[30] also identified obliquely descending eddies[14] with the vortex loops stretched in the braid, which envelop adjacent primary two-dimensional vortices (see Fig. 5.11). The vortex loops that initially emerge on a shear layer are obliquely stretched from the bottom part of a primary vortex toward the upper part of a successive primary vortex formed at the next plunging phase, leading to the rib structure. The inclination angle of the obliquely descending eddy (the vortex loop) increases from the first to the second plunging phase.

It has been shown through physical experiments of breaking waves that a planar overturning jet rapidly evolves into a finger-shaped secondary jet after wave plunging. The shape of the wave-breaking jets is formed by those longitudinal counter-rotating vortices that develop in regions of shear insta-

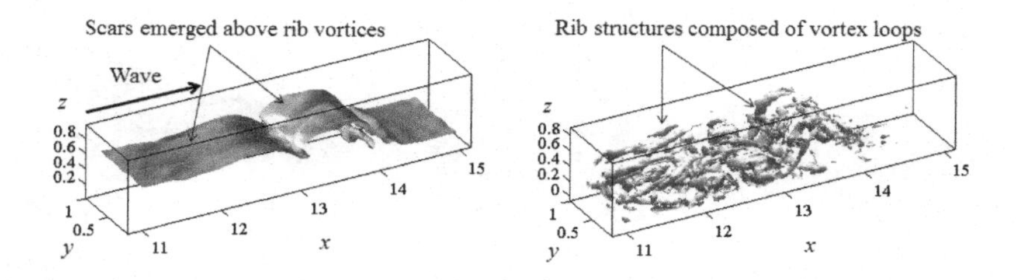

Fig. 5.10. Shape of the free-surface (left) and distribution of coherent vortex cores (right) in breaking waves.

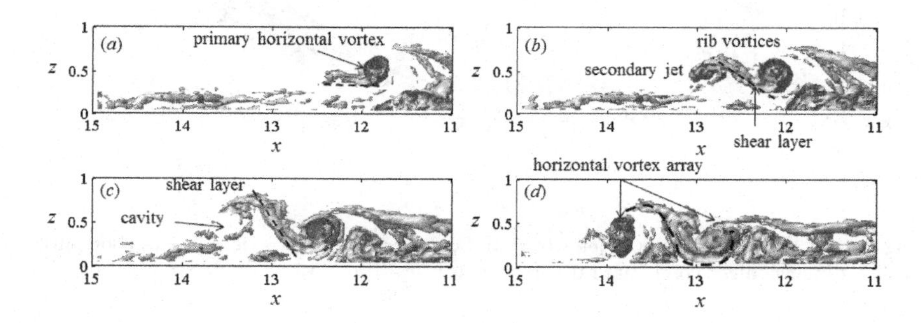

Fig. 5.11. Cross-sectional evolution of vortex cores after wave breaking; (*a*) the vortex loops initially emerge at stagnation region in front of the primary vortex at the first plunging point, (*b*) the rib structure composed of the vortex loops (obliquely descending eddy) are stretched and intensified on a shear layer between the secondary jet and primary vortex, (*c*) the inclination angle of the shear layer involving the ribs increases (second plunging point), (*d*) the front and rear edges of the ribs are wrapped by the adjacent primary horizontal vortices.

bility explained above. Saruwatari, Watanabe, and Ingram[31] computed the local surface deformation resulting from an oblique impact of a columnar water jet, using a three-dimensional large eddy simulation with the surface stress model described in §5.4.2 to model the overturning jet of a breaking wave.[22] The computation provides physically rationalized interpretations for the evolution of a splashing jet (see Fig. 5.12). As the secondary jet emerges, the vorticity field becomes unstable due to strong shear beneath the jet surface, and pairs of longitudinal counter-rotating vortices stretched along the direction of the jet projection are formed. The presence of these

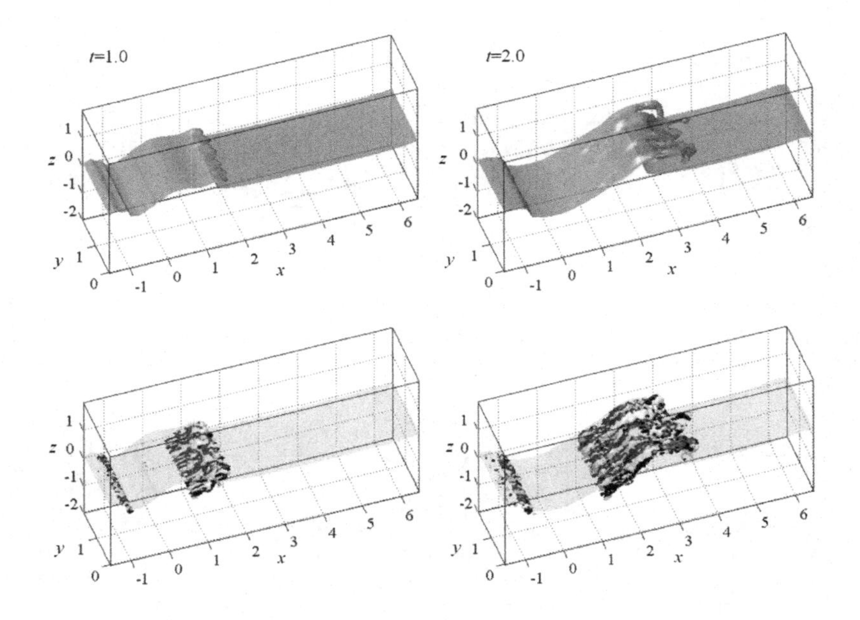

Fig. 5.12. Evolution of finger jets (top) and counter-rotating vortices in jets (bottom). white: dimensionless vorticity of 0.1, black: -0.1.

vortex pairs creates convergent surface flows to entrain the surface, resulting in the formation of longitudinal scars on the projecting jet. Following significant growth of the scars on both its upper and lower surfaces due to the surface entrainment by the vortices, the jet de-couples into fingers. As the jet fingers continue to be stretched along the jet axes, they become unstable and shed into droplets.

The three-dimensional vortices formed in breaking waves also have other important consequences in the coastal environment: sediment suspension and transport,[14] gas transfer via air bubbles entrappped within vortices,[30] and heat transfer through surface renewal by vortices,[32] processes that continue to be investigated in future computational research.

Okayasu *et al.*[17] and Suzuki *et al.*[23] also introduced the CIP method to a Smagorinsky type of LES (see details in §2.3.3) to model practical problems in coastal engineering: wave overtopping with breaking waves, and bottom shear and resulting sediment suspension in the surf zone (see Fig. 5.13). Kawasaki and Hakamata[6] developed a three-dimensional numerical wave tank to compute the CSF formulation for three-phase solid-

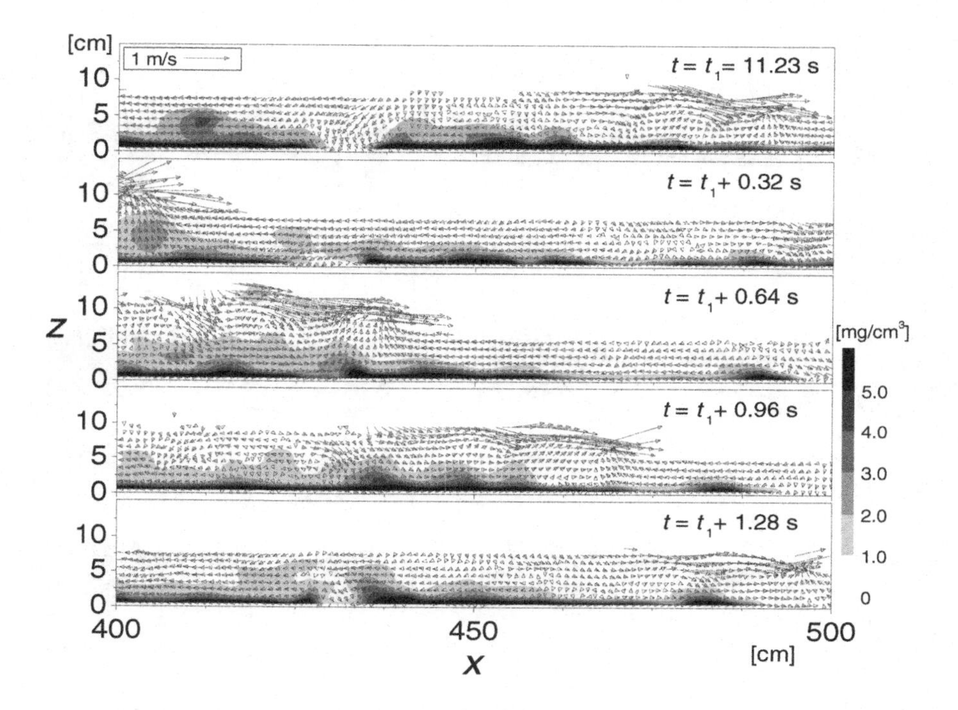

Fig. 5.13. Distributions of velocity and suspended sediment concentrations under breaking waves.[23]

gas-liquid flows using the CIPCLS method (§5.3.2). Realistic numerical experiments of floating bodies displaced by water waves generated by a numerical wave paddle can be performed using a boundary model for fluid-solid interactions (see Fig. 5.14).

Further practical applications of CIP methods to develop realistic numerical wave flumes are expected to accurately estimate coastal problems and design coastal structures.

Mutsuda *et al.*[12] used a dynamic model, explained in §2.3.5, to represent the unsteady turbulence that evolves at the two-phase air-water interface in breaking waves with a CCUP approach (§5.3.1). Mutsuda[13] extended his model to configure a numerical wind-wave tunnel to reproduce wind-wave evolution and white capping processes that result from wind shear. A typical turbulent boundary layer in the wind field blowing over a body of still water reasonably generates short-crested irregular waves (see Fig. 5.15). The computed three-dimensional wind flow involves

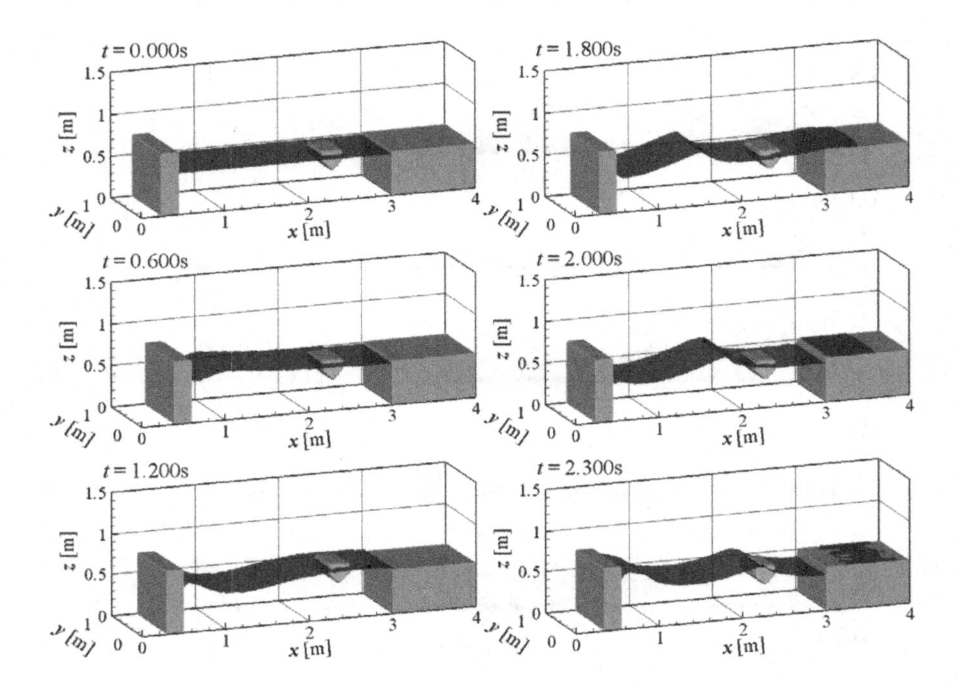

Fig. 5.14. Numerical wave basin illustrating floating bodies displaced by water waves generated by a numerical wave paddle.[6]

separation of the boundary layer from the wave crest, inducing undulations in the local air pressure and wind shear to develop waves. Air entrainment during white capping is also computed.

Extending the free-surface model explained in §5.4.2 to two-phase air-liquid flow configures another type of wind-wave tunnel, one which allows a sharp interface where the interfacial dynamic boundary condition Eq. (5.83) is reasonably satisfied.[33] Counter-rotating longitudinal vortices in the direction of the wind are produced in the wind boundary layer over the water surface, and shearing in the water flow is simultaneously driven underneath (see Fig. 5.16). A brick-pattern in the wind-wave surface is initially observed during the formation of this sheared interface.

Fluid flows within and across ocean-air surfaces significantly contribute to physical, chemical, and biological processes resulting from heat and gas transfers across ocean surfaces as well as momentum transfer from wind to ocean flows. In particular, huge amounts of air bubbles entrained into the water during white capping in the open ocean affects the oxygen saturation

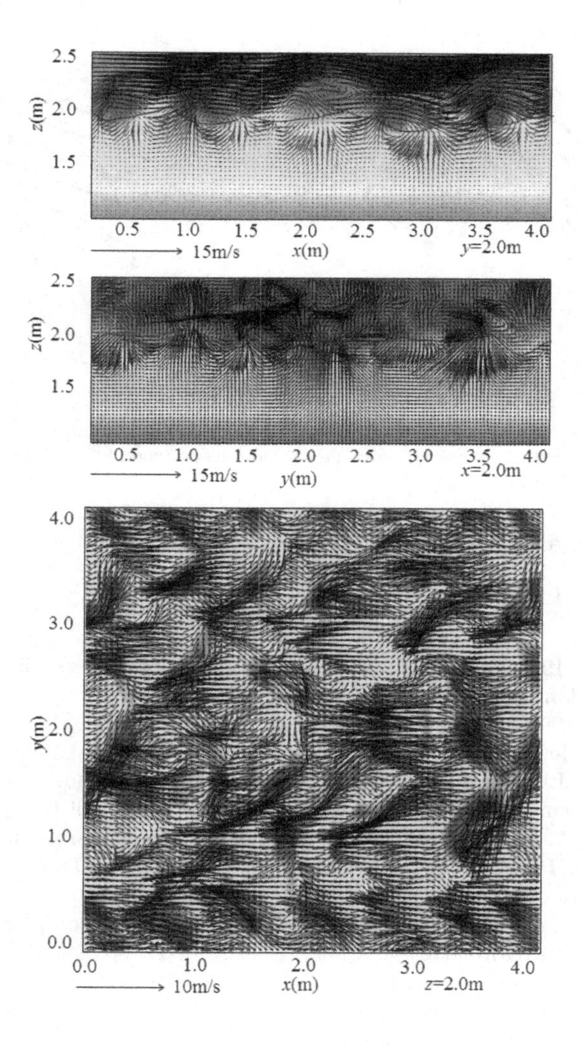

Fig. 5.15. Distributions of air and water velocities in the streamwise direction (top), spanwise direction (middle), and on the water surface (bottom) in a three-dimensional wind-wave tunnel.[13]

of seawater,[27] and sea sprays produced during wave breaking processes enhance heat and moisture exchanges between the atmosphere and the ocean, which often influences coastal weather and the regional climate.[9] Numerical wave computations are expected to further describe those interactions and complement parameterizations of atmospheric and ocean climate models,

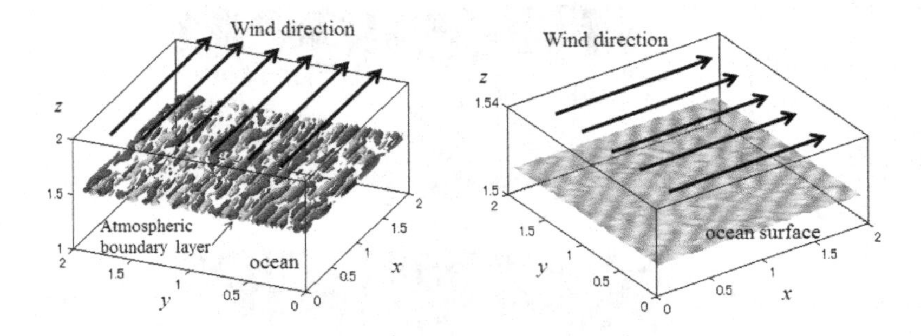

Fig. 5.16. Streamwise counter-rotating vortices in a wind boundary layer (left) and wind waves formed at an initial wave generation stage (right).[33] white: dimensionless vorticity of 10^{-4}, black: -10^{-4}.

which will be also discussed in §9.3.

References

1. Aoki T. (1995): Multi-dimensional advection of CIP (Cubic-Interpolated Propagation) scheme, *CFD Journal*, Vol.4, pp.279-291.
2. Aoki T. (1997): Interpolated Differential Operator (IDO) Scheme for solving partial differential equation, *Comput. Phys. Comm.*, Vol.102, pp.132-146.
3. Brackbill J.U., Kothe D.B. and C.Zemach (1992): A continuum method for modeling surface tension, *J. Computational Physics*, Vol.100, pp.335-354.
4. Esmaeeli A. and Tryggvason G. (1998): Direct numerical simulations of bubbly flows. Part1. Low Reynolds number arrays, *J. Fluid Mech.*, Vol.377, pp.313-345.
5. Hirt C.W. and Nichols B.D. (1980): Adding limited compressibility to incompressible hydrocodes, *J. Comp. Phys.*, Vol.34, pp.390-400.
6. Kawasaki K. and Hakamata M. (2007): Development of three-dimensional numerical model of multiphase flow "DILPHIN-3D" and dynamic analysis of drifting bodies under wave action, *Annu. J. Coastal Eng.*, JSCE, Vol.54, pp.31-35 (in Japanese).
7. Kelecy F.J. and Oletcher R.H. (1997): The development of a free surface capturing approach for multidimensional free surface flows in closed containers, *J. Comp. Phys.*, Vol.138, pp. 939-980.
8. Kim J. and Moin P. (1985): Application of a fractional-step method to incompressible Navier-Stokes equations, *J. Comp. Phys.*, Vol.59, pp.308-323.
9. De Leeuw G. (1999): Sea spray aerosol production from waves breaking in the surf zone. *J. Aerosol Sci.*, Vol. 30, Suppl. 1, S63-S64.
10. Longuet-Higgins M.S. (1992): Capillary rollers and bores, *J. Fluid Mech.*, Vol.240, pp.659-679.

11. Liow J.L. (2001): Splash formation by spherical drops, *J. Fluid Mech.*, Vol.427, pp.73-105.
12. Mutsuda H., Kawai H. and Yasuda T. (2000): Dynamic-LES of turbulence and bubble entrainment after wave breaking. *Proc. Coastal Eng., JSCE*, Vol.47, pp.171-175 (in Japanese).
13. Mutsuda H. (2001): Three-dimensional computation of wind-wave turbulent boundary layer flow with wave breaking, *Proc. Coastal Eng., JSCE*, Vol.48, pp.61-65 (in Japanese).
14. Nadaoka K., Hino M. and Koyano Y. (1989): Structure of the turbulent flow field under breaking waves in the surf zone, *J. Fluid Mech.*, Vol.204, pp.359-387.
15. Nakamura R., Tanaka T. and Yabe T. (2001): Multi-dimensional conservative scheme (CIP-CSL4) in non-conservative form, *CFD Journal*, pp.574-585.
16. Nakamura R., Tanaka T., Yabe T. and Takizawa K. (2001): Exactly conservative semi-Lagrangian scheme for multi-dimensional hyperbolic equations with directional splitting technique, *J. Comput. Phys.*, Vol.174, pp.171-207.
17. Okayasu A., Suzuki T. and Matsubayashi Y. (2005): Laboratory experiment and three-dimensional large eddy simulation of wave overtopping on gentle slope seawalls, *Coastal Eng. J.*, Vol.47, pp.71-89.
18. Osher S. and Sethian J. (1988): Fronts propagating with curvature dependent speed: Algorithms based on Hamilton-Jacobi formulations, *J.Comput.Phys.*, Vol.79, pp.12-49.
19. Osher S. and Fedkiw R.P. (2001): Level set methods: An overview and some recent results. *J. Comp. Phys.*, Vol.169, pp.463-502.
20. Popinet S. and Zaleski S. (2002): Bubble collapse near a solid boundary: a numerical study of the influence of viscosity, *J. Fluid Mech.*, Vol.464, pp.137-163.
21. Scardovelli R. and Zalenski S. (1999): Direct numerical simulation of free-surface and inter-facial flow, *Annu. Rev. Fluid Mech.*, Vol.31, pp.567-603.
22. Saruwatari A., Watanabe Y. and Ingram D.M. (2009): Scarifying and fingering surfaces of plunging jets. *Coastal Engineering*, Vol.56, pp.1109-1122.
23. Suzuki T., Okayasu A. and Shibayama T. (2007): A numerical study of intermittent sediment concentration under breaking waves in the surf zone, *Coastal Eng.*, Vol.54, pp.433-444.
24. Takewaki H., Nishiguchi A. and Yabe T. (1985): The Cubic-Interpolated Pseudo-Particle (CIP) Method for solving Hyperbolic-Type Equations, *J. Comp. Phys.*, Vol.61, pp.261-268.
25. Tanaka T., Nakamura R. and Yabe T. (2000): Constructing exactly conservative scheme in a non-conservative form, *Comput. Phys. Commu.* , Vol.126, pp.232-243.
26. Unverdi S.O. and Tryggvason G. (1992): A front-tracking method for viscous, incompressible, multi-fluid flows. *J. Comp. Phys.*, Vol.100, pp.25-37.
27. Wallace D.W.R. and Wirick C.D. (1992): Large air-sea gas fluxes associated with breaking waves. *Nature*, Vol.356, pp.694-696.
28. Watanabe Y. and Saeki H. (1999): Three-dimensional large eddy simulation of breaking waves, *Coastal Engineering Journal*, Vol.41, pp.281-301.

29. Watanabe Y. and Saeki H. (2002): Velocity field after wave breaking, *Int. J. Numer. Meth. Fluids*, Vol.39, pp.607-637.
30. Watanabe Y., Saeki H. and Hosking R.J. (2005): Three-dimensional vortex structures breaking waves, *J. Fluid Mech.*, Vol.545, pp.291-328.
31. Watanabe Y., Saruwatari A. and Ingram D.M. (2008): Free-surface flows under impacting droplets. *J. Comp. Phys.*, Vol.227, pp.2344-2365.
32. Watanabe Y. and Mori N. (2008): Infrared measurements of surface renewal and subsurface vortices in nearshore breaking waves. *J. Geophys. Res.*, Vol.113, C07015, doi:10.1029/2006JC003950.
33. Watanabe Y. and Iwashita A. (2011) Numerical scheme to fulfill interfacial dynamic boundary conditions, and its application to wind-wave generation (in Japanese). *J. JSCE, Ser. B2 (Coastal Engineering)*, Vol.67 (2), pp.16-20.
34. Yabe T. and Wang P.-Y. (1991): Unified numerical procedure for compressible and incompressible fluid. *J. Phys. Soc. Japan*, Vol.60 (7), pp.2105-2108.
35. Yabe T. and Wang P.-Y. (1991): A universal solver for hyperbolic equations by Cubic-Polynomial Interpolation I. One dimensional solver. *Comp. Phys. Comm.*, Vol.66, pp.219-232.
36. Yabe T. and Xiao F. (1993): Description of complex and sharp interface during shock wave interaction with liquid drop. *J. Phys. Soc. Japan*, Vol.62, pp.2537-2540.
37. Yabe T., Tanaka R., Nakamura T. and Xiao F. (2001): An exactly conservative semi-Lagrangian scheme (CIP-CSL) in one dimension, *Mon. Wea. Rev.*, Vol.129, pp.332-344.

Chapter 6

Particle Method

Hitoshi Gotoh

Since fluid dynamics and computations in a numerical wave flume are the primary topics of this book, this chapter details how the particle method can be applied to the Navier–Stokes equation (i.e., definition of the particle method in a narrow sense), explaining the mathematical model and focusing on the computational algorithms. Theoretical aspects of particle method are explained with taking care of integrating description of Smoothed Particle Hydrodynamics (SPH) method[54] and Moving Particle Semi-implicit (MPS) method.[48] The Sub Particle Scale (SPS)-turbulence model[30] as LES for particle method is introduced briefly. And the up-to-date accurate particle methods to attenuate unphysical pressure fluctuations are also explained in detail.

6.1. Concept of the Particle Method

A particle method means, in a broad sense, a mathematical model with multiple moving calculation points that interact with each other. In this sense, all of the following methods are particle methods: Smoothed Particle Hydrodynamics (SPH) method,[54] Moving Particle Semi-implicit (MPS) method,[48] Direct Simulation Monte Carlo (DSMC)[4] for molecular dynamics, Lattice Boltzmann Method (LBM)[56] for rarefied gas dynamics, in which a continuum approximation cannot be applied, and Distinct Element Method (DEM)[12] which has been applied to sediment transport dynamics as a granular material model.[19]

As a solver of Navier–Stokes equation, the particle method is completely different from the grid method, a method that fixes calculation points, or definition points of a physical property, on a computational grid. With the particle method, the Navier–Stokes equation is discretized with moving calculation points, capturing the interactions between particles.

Fluid motion can be described in two ways, namely Eulerian and Lagrangian. A particle method is Lagrangian when the particle or definition point moves with the fluid. The derivative D/Dt on the left side of Navier–Stokes equation is called the Lagrangian derivative (or substantial derivative), which consists of the time-derivative term and the advection term in an Eulerian format. In a Lagrangian format, the derivative D/Dt is equal to the time derivative term, and the advection term is unnecessary. In other words, in a Lagrangian format, the advection term is calculated as the movement of the calculation points.

The greatest difficulty with an Eulerian description is low resolution due to numerical diffusion associated with the discretization of the advection term. With the particle method, the advection term is unnecessary, and the computation is free from numerical diffusion.[51] This absence of numerical diffusion allows for superior tracking of the water surface using a simple algorithm and yielding high resolution. The particle method is especially suited for a topological change in the water surface, from being simply-connected to biconnected, an air-entrainment brought by plunging breaker. For example, when an impinging plunging jet generates a splash, the particle method is particularly applicable to model these violent flow motions in the surface water region.

Another advantage of the particle method is the ease with handling a boundary condition.[51] Because actual coastal structures have concave or convex geometries, such as lib and slit, arranging the computational grid along these complicated geometries is necessary in the grid method. In a three-dimensional field, grid generation is a complicated process that requires considerable skill. With the particle method, however, preprocessing consists of arranging particles along a fixed boundary, a simple task. In hydraulic engineering, for example, this advantage in preprocessing lends itself to computing the flow through a fish ladder with a complicated arrangement of basins.[31]

There are two major particle methods: the SPH (Smoothed Particle Hydrodynamics) method[54] and the MPS (Moving Particle Semi-implicit) method.[48] The SPH method has its origin in astrophysics; it has been applied to model the expansion of the universe after the Big Bang.

In the prototypical SPH method, a compressible fluid is computed with an explicit algorithm. The MPS method, however, uses a semi-implicit method for modeling an incompressible fluid such as water.[48] Both methods assume the continuum approximation of a fluid without using a computational grid. Additionally, because these particle methods can be applied to the dynamics of continuum, both fluids and elastic bodies can be modeled.[51] Their application spans many fields, such as mechanical engineering (nuclear engineering and naval architecture, etc.) or civil engineering (hydraulic engineering, structural engineering and geotechnical engineering, etc.).

6.2. Discretization of Governing Equations

As explained in §1.1, the motion of an incompressible viscous fluid is described by the Navier–Stokes and continuity equations:

$$\left.\begin{aligned} &\frac{1}{\rho}\frac{D\rho}{Dt} + \boldsymbol{\nabla}\cdot\boldsymbol{u} = 0 \\ &\frac{D\boldsymbol{u}}{Dt} = -\frac{1}{\rho}\boldsymbol{\nabla}p + \boldsymbol{g} + \nu\boldsymbol{\nabla}^2\boldsymbol{u} \end{aligned}\right\} \tag{6.1}$$

where $\boldsymbol{u}$ is the velocity vector, t is time, ρ is the fluid density, p is the pressure, $\boldsymbol{g}$ is the gravitational acceleration vector and ν is the laminar kinematic viscosity. It should be noted that Eq. (6.1) is written for a compressible fluid. Incompressibility is enforced by setting $D\rho/Dt = 0$ at each particle for each calculation time step. The left hand side of the momentum equation, namely Eq. (6.1), denotes the substantial derivative involving the advection term. In both the SPH and MPS methods, the advection term is automatically calculated through the tracking of particle motion. Hence, numerical diffusion arising from successive interpolation of the advection terms in Eulerian grid-based methods is well controlled without a sophisticated algorithm.

In particle methods the fluid is regarded as a finite number of particles, each possessing its own mass and other physical properties. To solve the Navier–Stokes and continuity equations, spatial derivatives, or vector derivative operators, must be calculated. The operation for calculating spatial derivatives of physical properties is called integral interpolation. The difference between the SPH and MPS methods is characterized by the integral interpolants. In the next section, integral interpolants and vector derivative operators of the SPH and MPS methods will be explained.

6.2.1. *Integral interpolants of the SPH method*

In the SPH method, the integral interpolants provides an approximation of an arbitrary physical property $\phi(\boldsymbol{x})$ in terms of a kernel function $W(|\boldsymbol{r}|, h)$

as follows:

$$\phi(\boldsymbol{x}) = \iiint_{\Omega} \phi(\boldsymbol{\xi})\, W\left(|\boldsymbol{r}|, h\right)\, dV \quad ; \quad \boldsymbol{r} = \boldsymbol{\xi} - \boldsymbol{x} \tag{6.2}$$

where h is the smoothing length taken slightly larger than the particle diameter, and $\boldsymbol{x}$ and $\boldsymbol{\xi}$ are position vectors of target particle and its neighboring particle, respectively. The kernel function should have the following basic properties:

$$W\left(|\boldsymbol{r}|, h\right) \geq 0 \quad \text{for} \quad \xi \in \Omega \tag{6.3}$$

$$\iiint_{\Omega} W\left(|\boldsymbol{r}|, h\right) dV = 1 \tag{6.4}$$

$$\nabla W\left(|\boldsymbol{r}|, h\right)|_{x} = -\nabla W\left(|\boldsymbol{r}|, h\right)|_{\xi} \tag{6.5}$$

In an SPH-based calculation, the kernel function not only affects CPU time, but also the stability of the calculation. When the second derivative of a kernel function is discontinuous, it becomes quite sensitive to particle disorders or randomness of particle positions,[59] and instability may arise. Most SPH calculations employ the cubic B-spline kernel proposed by Monaghan:[59]

$$W(Q, h) = \begin{cases} \dfrac{10}{7\pi h^2}\left(1 - \dfrac{3}{2}Q^2 + \dfrac{3}{4}Q^3\right) & Q < 1 \\ \dfrac{10}{28\pi h^2}(2 - Q)^3 & 1 \leq Q \leq 2 \\ 0 & Q > 2 \end{cases} \tag{6.6}$$

where

$$Q = \frac{|\boldsymbol{r}|}{h} \tag{6.7}$$

The cubic B-spline kernel resembles a Gaussian kernel while maintaining compact support, meaning the interactions are exactly zero for $|\boldsymbol{r}| > 2h$. In addition, the second derivative of a cubic B-Spline kernel is continuous, and the dominant error term in the integral interpolant is $O(h^2)$.[59] Morris *et al.*[64] applied a higher-order Quintic spline kernel:

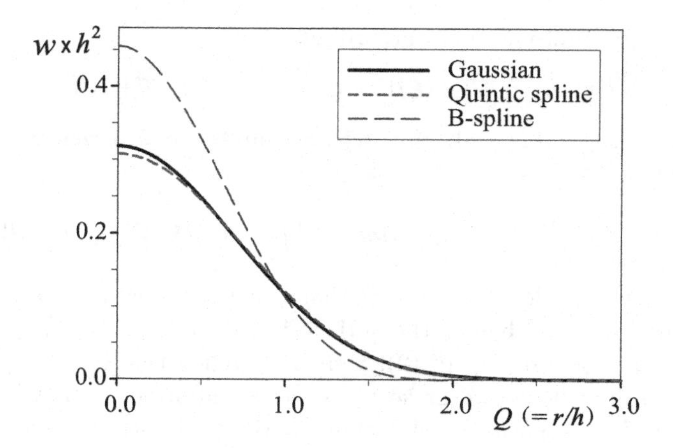

Fig. 6.1. Common kernels of the SPH method.

$$W(Q,h) = \begin{cases} \dfrac{7}{478\pi h^2}\left[(3-Q)^5 - 6(2-Q)^5 + 15(1-Q)^5\right] & Q < 1 \\ \dfrac{7}{478\pi h^2}\left[(3-Q)^5 - 6(2-Q)^5\right] & 1 \le Q < 2 \\ \dfrac{7}{478\pi h^2}\left[(3-Q)^5\right] & 2 \le Q \le 3 \\ 0 & Q > 3 \end{cases} \tag{6.8}$$

The second derivative of the above kernel is of order 3. As a result, calculations would be less sensitive to particle disorders. However, in comparison to the B-Spline kernel, the Quintic spline kernel requires longer computational time. Figure 6.1 presents a sketch of the most common kernels applied in an SPH framework. Alternative kernels in SPH simulations include a modified Gaussian[10] and Quadratic[13] kernels.

Based on Eq. (6.2), the particle-based approximation for the derivative of a scalar function can be deduced by

$$\nabla\phi(\boldsymbol{x})|_x = \iiint_\Omega \nabla\phi(\boldsymbol{\xi})|_\xi \, W(|\boldsymbol{r}|, h)\, dV \tag{6.9}$$

where $\nabla\phi(\boldsymbol{x})|_x$ denotes the gradient of $\phi(\boldsymbol{x})$ at particle position $\boldsymbol{x}$. The following simple identity is used to derive the SPH integral interpolant:

$$\nabla\left\{\phi(\boldsymbol{\xi})W(|\boldsymbol{r}|, h)\right\}|_\xi = \nabla\phi(\boldsymbol{\xi})|_\xi \, W(|\boldsymbol{r}|, h) + \phi(\boldsymbol{\xi})\, \nabla W(|\boldsymbol{r}|, h)|_\xi \tag{6.10}$$

Taking Eq. (6.5) into account, accordingly,

$$\nabla\phi(\boldsymbol{\xi})|_{\xi}\, W(|\boldsymbol{r}|,h) = \nabla\{\phi(\boldsymbol{\xi})W(|\boldsymbol{r}|,h)\}|_{\xi} + \phi(\boldsymbol{\xi})\,\nabla W(|\boldsymbol{r}|,h)|_{x} \quad (6.11)$$

Considering Eq. (6.11) and applying the Gauss' (or divergence) theorem, Eq. (6.9) is written as

$$\nabla\phi(\boldsymbol{x})|_{x} = \iint_{S}\phi(\boldsymbol{\xi})W(|\boldsymbol{r}|,h)\,\boldsymbol{n}dS + \iiint_{\Omega}\phi(\boldsymbol{\xi})\,\nabla W(|\boldsymbol{r}|,h)|_{x}dV \quad (6.12)$$

In astrophysics problems, we assume that either the function ϕ or the kernel is zero on surface S; hence, the surface integration, or the first term on the right-hand side of Eq. (6.12), vanishes. When the SPH is applied to free-surface fluid flows, we make the same assumption, although such an assumption is not entirely valid close to the free surface. Ignoring the surface term, the spatial derivative of a function is obtained by simply shifting the ∇ operator from the physical quantity to the kernel function.

As a result, the gradient operator is written as

$$\nabla\phi(\boldsymbol{x})|_{x} = \iiint_{\Omega}\phi(\boldsymbol{\xi})\;\nabla W(|\boldsymbol{r}|,h)|_{x}dV \quad (6.13)$$

Similarly, the divergence of a vector quantity is approximated as

$$\nabla\cdot\boldsymbol{\phi}(\boldsymbol{x})\,|_{x} = \iiint_{\Omega}\boldsymbol{\phi}(\boldsymbol{\xi})\cdot\nabla W(|\boldsymbol{r}|,h)|_{x}dV \quad (6.14)$$

With particle methods, computational information is known solely at discrete points. As such, integrals are evaluated as sums over neighboring particles

$$\left.\begin{aligned} \phi(\boldsymbol{x}_i) &= \sum_{j\neq i}\phi(\boldsymbol{x}_j)\,W(|\boldsymbol{x}_j-\boldsymbol{x}_i|,h)\,V_j \\ V_j &= \frac{m_j}{\rho_j} = \frac{1}{\sum\limits_{j\neq i}^{M} W(|\boldsymbol{x}_j-\boldsymbol{x}_i|,h)} \end{aligned}\right\} \quad (6.15)$$

where m_j, ρ_j, and V_j are the mass, density, and tributary (or statistical) volume associated with neighboring particle j, respectively. The integration is calculated for each particle in a circular (or spherical) domain or influence circle (or sphere) of particle i with radius r_e. Accordingly, the gradient of a scalar quantity and divergence of a vector quantity are approximated by the following equations:

$$\phi(\boldsymbol{x}_i) = \sum_{j\neq i}\phi(\boldsymbol{x}_j)\nabla W\;(|\boldsymbol{x}_j-\boldsymbol{x}_i|,h)|_{x_i}\,V_j \quad (6.16)$$

$$\nabla \cdot \boldsymbol{\phi}(\boldsymbol{x}_i) = \sum_{j \neq i} \boldsymbol{\phi}(\boldsymbol{x}_j) \cdot \nabla W\left(\left|\boldsymbol{x}_j - \boldsymbol{x}_i\right|, h\right)\Big|_{\boldsymbol{x}_i} V_j \tag{6.17}$$

6.2.2. *Vector differential operators of the SPH method*

In the SPH method, the density of particle i is obtained by summing over the contributions of the neighboring particles:

$$\rho_i = \sum_{j \neq i}^{M} m_j W_{ij} \quad ; \quad W_{ij} = W\left(\left|\boldsymbol{x}_j - \boldsymbol{x}_i\right|, h\right) \tag{6.18}$$

Considering Eq. (6.13), the pressure gradient on particle i is expressed as

$$-\left.\frac{\nabla p}{\rho}\right|_i = -\frac{1}{\rho_i} \sum_{j \neq i} \frac{m_j}{\rho_j} p_j \left.\nabla W_{ij}\right|_i \tag{6.19}$$

However, this equation does not conserve either linear or angular momentum since the force on particle i due to j is not equal and opposite to the force on particle j due to i. In equation form,

$$\frac{m_i m_j p_j}{\rho_i \rho_j} \left.\nabla W_{ij}\right|_i \neq -\frac{m_i m_j p_i}{\rho_i \rho_j} \left.\nabla W_{ij}\right|_i \tag{6.20}$$

Note that $\left.\nabla W_{ij}\right|_i = -\left.\nabla W_{ij}\right|_j$. Eq. (6.19) can be written in a conservative form by placing density inside the operator (Monaghan[59]):

$$\nabla\left(\frac{p}{\rho}\right) = \frac{\rho \nabla p - p \nabla \rho}{\rho^2} = \frac{\nabla p}{\rho} - \frac{p}{\rho^2} \nabla \rho \tag{6.21}$$

Hence,

$$\begin{aligned} -\left.\frac{\nabla p}{\rho}\right|_i &= -\sum_{j \neq i} \frac{m_j}{\rho_j} \frac{p_j}{\rho_j} \left.\nabla W_{ij}\right|_i - \sum_{j \neq i} \frac{p_i}{\rho_i^2} m_j \left.\nabla W_{ij}\right|_i \\ &= -\sum_{j \neq i} m_j \left(\frac{p_j}{\rho_j^2} + \frac{p_i}{\rho_i^2}\right) \left.\nabla W_{ij}\right|_i \end{aligned} \tag{6.22}$$

The commonly applied Laplacian model in an SPH framework is founded on a Taylor series expansion of a physical quantity, ϕ_i, with respect to the physical quantity at a neighboring particle j, ϕ_j. Neglecting the second order terms,

$$\phi_i = \phi_j + \nabla \phi_j \cdot (\boldsymbol{x}_i - \boldsymbol{x}_j) \tag{6.23}$$

The Laplacian of a physical quantity ϕ at a target particle i can be derived by taking the divergence of the gradient model and by considering a simplified gradient approximation founded on the above mentioned Taylor series expansion.

$$\nabla\phi_j = (\phi_j - \phi_i)\frac{\boldsymbol{x}_j - \boldsymbol{x}_i}{|\boldsymbol{x}_j - \boldsymbol{x}_i|^2} \tag{6.24}$$

$$\begin{aligned}\nabla^2\phi\big|_i = \nabla\cdot(\nabla\phi)\big|_i &= \sum_{j\neq i}\frac{m_j}{\rho_j}\nabla\phi_j\cdot\nabla W_{ij}\big|_i \\ &= \sum_{j\neq i}\frac{m_j}{\rho_j}(\phi_j - \phi_i)\frac{(\boldsymbol{x}_j - \boldsymbol{x}_i)\cdot\nabla W_{ij}\big|_i}{|\boldsymbol{x}_j - \boldsymbol{x}_i|^2}\end{aligned} \tag{6.25}$$

Placing the density inside the divergence operator,

$$\nabla\cdot\left(\frac{\nabla\phi}{\rho}\right) = \nabla\left(\frac{1}{\rho}\right)\cdot\nabla\phi + \frac{1}{\rho}\nabla^2\phi \tag{6.26}$$

and considering the following two approximations,

$$\nabla\phi\big|_i = (\phi_i - \phi_j)\frac{\boldsymbol{x}_i - \boldsymbol{x}_j}{|\boldsymbol{x}_i - \boldsymbol{x}_j|^2} \tag{6.27}$$

$$\nabla\left(\frac{1}{\rho}\right)\bigg|_i = \sum_{j\neq i}\frac{m_j}{\rho_j}\frac{1}{\rho_j}\nabla W_{ij}\big|_i \tag{6.28}$$

the SPH Laplacian model is derived as:

$$\nabla\cdot\left(\frac{\nabla\phi}{\rho}\right)\bigg|_i = \sum_{j\neq i}\frac{m_j}{\rho_j}\left(\frac{1}{\rho_j} + \frac{1}{\rho_i}\right)(\phi_j - \phi_i)\frac{(\boldsymbol{x}_j - \boldsymbol{x}_i)\cdot\nabla W_{ij}\big|_i}{|\boldsymbol{x}_j - \boldsymbol{x}_i|^2} \tag{6.29}$$

Considering the wide range of densities of the particles close to the free surface, the calculated densities at target particle i and neighboring particle j are simply replaced by their arithmetic averages

$$\rho_i \to \frac{\rho_i + \rho_j}{2} \quad ; \quad \rho_j \to \frac{\rho_i + \rho_j}{2} \tag{6.30}$$

and the symmetric formulation

$$\nabla\cdot\left(\frac{\nabla\phi}{\rho}\right)\bigg|_i = \sum_{j\neq i}\frac{8m_j}{(\rho_i + \rho_j)^2}(\phi_j - \phi_i)\frac{(\boldsymbol{x}_j - \boldsymbol{x}_i)\cdot\nabla W_{ij}\big|_i}{|\boldsymbol{x}_j - \boldsymbol{x}_i|^2} \tag{6.31}$$

is derived (Shao and Lo[69]).

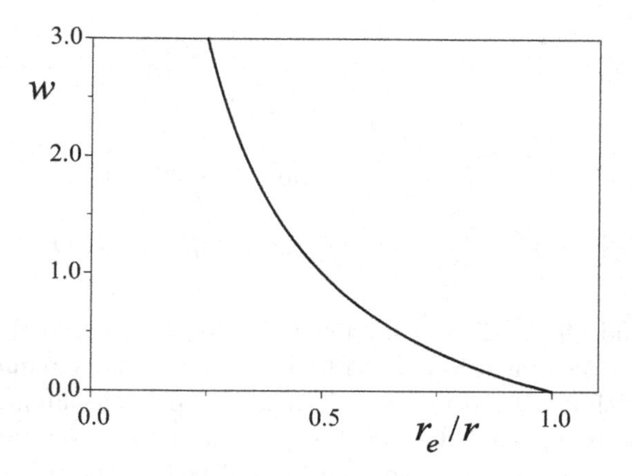

Fig. 6.2. Standard MPS kernel.

6.2.3. *Integral interpolants and vector differential operators of the MPS method*

Koshizuka and Oka[48] originally proposed the MPS method to simulate incompressible fluid flows. This method is similar to the SPH method in many aspects, including the kernel-based approximation by integral interpolants. Nevertheless, the MPS method employs a more simplified and comprehensible approach to approximate the gradient and Laplacian.

The commonly applied kernel function in the MPS method is defined as:

$$w(r) = \begin{cases} \dfrac{r_e}{r} - 1 & 0 \le r < r_e \\ 0 & r_e \le r \end{cases} \tag{6.32}$$

$$r = |\boldsymbol{x}_j - \boldsymbol{x}_i| \tag{6.33}$$

where r_e is the radius of influence. The above kernel function has two important features. First, the function is infinite at $r = 0$. As a result, particle clustering would be avoided. Second, the kernel is infinitely differentiable. The standard MPS kernel is illustrated in Fig. 6.2. In the MPS method, the statistical volume associated with particle j is denoted by the so-called particle number density, or n, defined as

$$n_i = \sum_{i \neq j} w\left(|\boldsymbol{x}_j - \boldsymbol{x}_i|\right) \tag{6.34}$$

Considering the following first order approximation,

$$\nabla \phi_{ij} = \frac{\phi_j - \phi_i}{|\boldsymbol{x}_j - \boldsymbol{x}_i|^2} (\boldsymbol{x}_j - \boldsymbol{x}_i) \tag{6.35}$$

the gradient at target particle i is simply expressed as:

$$\nabla \phi|_i = \frac{1}{n_i} \sum_{i \neq j} \frac{\phi_j - \phi_i}{|\boldsymbol{x}_j - \boldsymbol{x}_i|^2} (\boldsymbol{x}_j - \boldsymbol{x}_i) w(|\boldsymbol{x}_j - \boldsymbol{x}_i|) \tag{6.36}$$

Koshizuka and Oka[48] slightly modified the above equation to account for the effect of space dimensionality. Fluid incompressibility requires that the fluid density should be constant. Because the particle number density is proportional to the fluid density, the particle number density should be constant. Hence n_i in Eq. (6.36) can be replaced by the reference particle number density n_0. Accordingly, the MPS gradient and divergence models are respectively expressed as

$$\nabla \phi|_i = \frac{D_s}{n_0} \sum_{i \neq j} \frac{\phi_j - \phi_i}{|\boldsymbol{x}_j - \boldsymbol{x}_i|^2} (\boldsymbol{x}_j - \boldsymbol{x}_i) w(|\boldsymbol{x}_j - \boldsymbol{x}_i|) \tag{6.37}$$

and

$$\nabla \cdot \boldsymbol{\phi}|_i = \frac{D_s}{n_0} \sum_{i \neq j} \frac{(\boldsymbol{\phi}_j - \boldsymbol{\phi}_i) \cdot (\boldsymbol{x}_j - \boldsymbol{x}_i)}{|\boldsymbol{x}_j - \boldsymbol{x}_i|^2} w(|\boldsymbol{x}_j - \boldsymbol{x}_i|) \tag{6.38}$$

where D_s is the number of the spatial dimension. Despite its simplicity and being only of first order accuracy, the MPS gradient model has a distinct advantage over that of the SPH; that is, the sources of numerical errors tend to be more clearly known, with the exception of the surface integral in the SPH gradient approximation of Eq. (6.12).

The MPS Laplacian model is also derived based on a simple, comprehensible context. In spherical geometry, the Laplacian is expressed as

$$\nabla^2 \phi = \frac{1}{r^2} \frac{\partial}{\partial r} \left(r^2 \frac{\partial \phi}{\partial r} \right) = \frac{1}{r^2} \left(2r \frac{\partial \phi}{\partial r} + r^2 \frac{\partial^2 \phi}{\partial r^2} \right) \tag{6.39}$$

A Taylor series expansion up to second order differentiation yields

$$r_{ij}^2 \frac{\partial^2 \phi_{ij}}{\partial r_{ij}^2} = 2\phi_{ij} - 2r_{ij} \frac{\partial \phi_{ij}}{\partial r_{ij}} \tag{6.40}$$

where

$$\phi_{ij} = \phi_j - \phi_i \quad ; \quad r_{ij} = r_j - r_i \tag{6.41}$$

When combined with Eq. (6.39), we obtain

$$\nabla^2 \phi_{ij} = \frac{1}{r_{ij}^2} \frac{\partial}{\partial r_{ij}} \left(r_{ij}^2 \frac{\partial \phi_{ij}}{\partial r_{ij}} \right) = \frac{2\phi_{ij}}{r_{ij}^2} \tag{6.42}$$

Considering the definitions of r_{ij} and ϕ_{ij}, Eq. (6.42) is rewritten as

$$\nabla^2 \phi_{ij} = \frac{2\phi_{ij}}{r_{ij}^2} = \frac{2(\phi_j - \phi_i)}{|\boldsymbol{x}_j - \boldsymbol{x}_i|^2} \tag{6.43}$$

Accordingly, the Laplacian at a target particle i is obtained by a local weighted averaging of approximated Laplacians at neighboring particles j, i.e.,

$$\nabla^2 \phi\Big|_i = \frac{2}{n_0} \sum_{i \neq j} \frac{(\phi_j - \phi_i)}{|\boldsymbol{x}_j - \boldsymbol{x}_i|^2} w(|\boldsymbol{x}_j - \boldsymbol{x}_i|) \tag{6.44}$$

Similarly, the effect of space dimensionality is taken into account with the variable D_s, resulting in

$$\nabla^2 \phi\Big|_i = \frac{2D_s}{n_0} \sum_{i \neq j} \frac{(\phi_j - \phi_i)}{|\boldsymbol{x}_j - \boldsymbol{x}_i|^2} w(|\boldsymbol{x}_j - \boldsymbol{x}_i|) \tag{6.45}$$

The above equation is further modified to equate the increase of variance due to the re-distribution of the MPS Laplacian model with the increase of variance estimated from the unsteady diffusion equation:[48]

$$\nabla^2 \phi\Big|_i = \frac{2D_s}{\lambda n_0} \sum_{i \neq j} (\phi_j - \phi_i) w(|\boldsymbol{x}_j - \boldsymbol{x}_i|) \tag{6.46}$$

where

$$\lambda = \frac{\sum_{j \neq i} |\boldsymbol{x}_j - \boldsymbol{x}_i|^2 w(|\boldsymbol{x}_j - \boldsymbol{x}_i|)}{\sum_{j \neq i} w(|\boldsymbol{x}_j - \boldsymbol{x}_i|)} \tag{6.47}$$

6.3. Algorithms of Particle Methods

There are two major algorithms to solve the discretized equations of particle methods. The conventional simplified approach treats the fluid as weakly (slightly) compressible and calculates the pressure explicitly from an appropriate equation of state. This approach is usually known as the Weakly Compressible SPH (WCSPH) method. On the other hand, the more favored, physically sound and mathematically rigorous approach is to apply the so-called projection method. This approach results in a Poisson Pressure Equation (PPE) and an implicit calculation of pressure. The MPS method is founded on this projection method and hence its solution process is a semi-implicit one. By applying the same approach, Shao and Lo[69] proposed a semi-implicit version of the SPH method denoted as the Incompressible SPH (ISPH) method.

6.3.1. *The WCSPH method (fully explicit algorithm)*

Following the development of the SPH method by Lucy,[54] Monaghan[59] pioneered the first application of the SPH method to simulate fluid flows. Later, Monaghan[60] published a paper on the simulation of incompressible fluid flows by incorporating a modified equation of state relating the fluid pressure with the local density

$$p = B\left\{\left(\frac{\rho}{\rho_0}\right)^{\gamma} - 1\right\} \quad ; \quad \gamma = 7.0 \tag{6.48}$$

where ρ_0 is a reference density for the fluid. Here, B is a set parameter such that the speed of sound is:

$$C_s = \sqrt{\frac{\partial p}{\partial \rho}} \approx 10\, u_{max} \tag{6.49}$$

where u_{max} is the maximum flow velocity in the calculations. The actual speed of sound is not used since a small time step is required for numerical stability. Since the Mach number $M(= u_{max}/C_s) = 0.1$ and compressibility effects are $O(M^2)$, theoretically, the changes in fluid density would be within 1%. However, under particular flow conditions, such as those characterized with stagnation points, a further reduction in the Mach number (and therefore the time step) may be necessary to keep the density fluctuations within 1%.[15]

In addition to Eq. (6.48), the WCSPH approach considers the following governing equations:

$$\frac{d\boldsymbol{u}_i}{dt} = -\sum_{j \neq i} m_j \left(\frac{p_j}{\rho_j^2} + \frac{p_i}{\rho_i^2} + \Pi_{ij}\right) \nabla W_{ij}\big|_i \tag{6.50}$$

$$\frac{d\boldsymbol{x}_i}{dt} = \boldsymbol{u}_i \tag{6.51}$$

where $\boldsymbol{u}_i$ and $\boldsymbol{x}_i$ denote the velocity and position vectors corresponding to target particle i. In Eq. (6.50), the term Π_{ij} is the so-called artificial viscosity originally proposed to stabilize the calculations. The most frequently used artificial viscosity is the one proposed by Monaghan[59]

$$\Pi_{ij} = \begin{cases} \dfrac{-\alpha\overline{C}_{sij}\mu_{ij} + \beta\mu_{ij}^2}{\overline{\rho}_{ij}} & \boldsymbol{u}_{ij}\cdot\boldsymbol{x}_{ij} < 0 \\ 0 & \boldsymbol{u}_{ij}\cdot\boldsymbol{x}_{ij} \geq 0 \end{cases} \tag{6.52}$$

where

$$\boldsymbol{u}_{ij} = \boldsymbol{u}_j - \boldsymbol{u}_i \quad ; \quad \boldsymbol{x}_{ij} = \boldsymbol{x}_j - \boldsymbol{x}_i \tag{6.53}$$

$$\overline{C}_{sij} = \frac{C_{si} + C_{sj}}{2} \quad ; \quad \overline{\rho}_{ij} = \frac{\rho_i + \rho_j}{2} \tag{6.54}$$

$$\mu_{ij} = \frac{h\boldsymbol{u}_{ij}\cdot\boldsymbol{r}_{ij}}{r_{ij}^2 + 0.01h^2} \tag{6.55}$$

There are a few advantages and disadvantages associated with the artificial viscosity term. In addition to being a Galilean invariant, this type of viscosity term conserves both linear and angular momentum and vanishes for rigid body rotations (Monaghan[59]). However, it is a scalar viscosity and cannot account for flow direction. In addition, strong dissipation occurs in simulations with complex shearing, large vorticity decay, and unphysical momentum transfer (Ellero *et al.*[15]). More importantly, improperly selecting the empirical coefficients α and β in Eq. (6.52) results in additional unreal dissipations. Hence, in lieu of applying an artificial viscosity term, it is preferable to model viscosity in a realistic manner and keep the calculations stable by applying accurate and consistent numerical schemes.

6.3.2. *The MPS method (semi-implicit algorithm)*

As a projection-based particle method, the MPS method applies a Helmholtz–Hodge decomposition to the vector field. Accordingly, the velocity vector at time step $k+1$, $\boldsymbol{u}_{k+1}$, consists of the velocity vector at time step k, $\boldsymbol{u}_k$, plus two other incremental velocity vectors corresponding to the prediction and correction steps of pseudo time step $k+1/2$, i.e.,

$$\boldsymbol{u}_{k+1} = \boldsymbol{u}_k + \Delta\boldsymbol{u}_k^* + \Delta\boldsymbol{u}_k^{**} \tag{6.56}$$

The prediction step is an explicit step to predict the velocity of particles under given viscosity and gravity terms. The corresponding increment in the velocity field is written as

$$\Delta \boldsymbol{u}_k^* = \left(\nu \boldsymbol{\nabla}^2 \boldsymbol{u}\right)_k \Delta t + \boldsymbol{g} \Delta t \tag{6.57}$$

Consequently, the updated velocity and position vectors are given in the following forms:

$$\boldsymbol{u}_k^* = \boldsymbol{u}_k + \Delta \boldsymbol{u}_k^* \quad ; \quad \boldsymbol{x}_k^* = \boldsymbol{x}_k + \Delta \boldsymbol{u}_k^* \Delta t \tag{6.58}$$

In the temporal field, which is calculated in the prediction step, the volume conservation, namely the conservation of particle number density, is not satisfied. Therefore, in the correction step, the particle number densities must be corrected so that the remaining volume is conserved:

$$n_k^* + \Delta n_k^{**} = n_0 \tag{6.59}$$

The correction of particle number densities is achieved by updating the pressure calculation corresponding to time step $k+1$. The incremental velocity vector of the correction step is written as

$$\Delta \boldsymbol{u}_k^{**} = -\frac{1}{\rho} \boldsymbol{\nabla} p_{k+1} \Delta t \tag{6.60}$$

The updated pressure, p_{k+1}, required to maintain the volume conservation is calculated with a PPE. The PPE is obtained by considering the mass conservation equation

$$\frac{D\rho}{Dt} + \rho_0 \boldsymbol{\nabla} \cdot \boldsymbol{u} = 0 \tag{6.61}$$

which is written as follows in the MPS approach:

$$\frac{1}{n_0} \frac{Dn}{Dt} + \boldsymbol{\nabla} \cdot \boldsymbol{u} = 0 \tag{6.62}$$

The velocities and particle number densities in the correction process satisfy the mass conservation law

$$\frac{1}{n_0} \frac{\Delta n_k^{**}}{\Delta t} + \boldsymbol{\nabla} \cdot \Delta \boldsymbol{u}_k^{**} = 0 \tag{6.63}$$

Accordingly, and by considering Eq. (6.60), the PPE is derived as

$$\boldsymbol{\nabla}^2 p_{k+1} = \frac{\rho_0}{\Delta t^2} \frac{n_k^* - n_0}{n_0} \tag{6.64}$$

By solving Eq. (6.64) implicitly, the pressure is calculated; then the velocity correction vector at the correction step can be calculated with Eq. (6.60). Thus, the updated position vector, $\boldsymbol{x}_{k+1}$, is obtained as

$$\boldsymbol{x}_{k+1} = \boldsymbol{x}_k^* + \Delta \boldsymbol{u}_k^{**} \Delta t \tag{6.65}$$

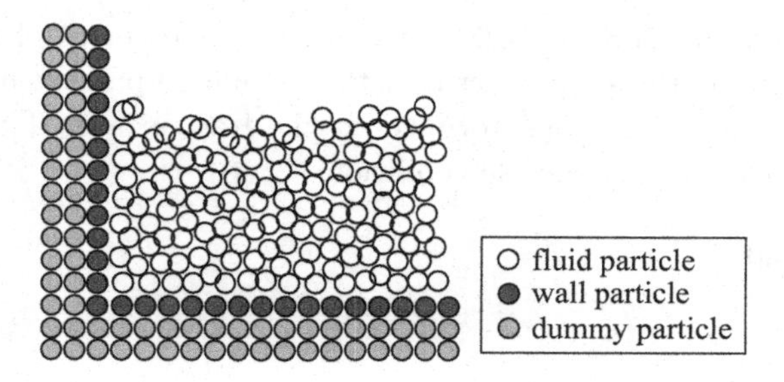

Fig. 6.3. Wall boundaries of the particle method.

The calculation time step increment is set according to the Courant stability condition and a maximum time resolution.

$$\Delta t = \min\left(\alpha_{dt} d_0 / u_{\max},\ 1.0 \times 10^{-3}\right) \tag{6.66}$$

To ensure the stability of MPS-based simulations, the pressure gradient is defined by replacing p_i with the minimum pressure value among its neighboring particles, $\hat{p}_i$, such as:

$$\nabla p|_i = \frac{D_s}{n_0} \sum_{j \neq i} \frac{p_j - \hat{p}_i}{|\boldsymbol{x}_j - \boldsymbol{x}_i|^2} (\boldsymbol{x}_j - \boldsymbol{x}_i) w\left(|\boldsymbol{x}_j - \boldsymbol{x}_i|\right) \tag{6.67}$$

$$\hat{p}_i = \min_{j \in J}(p_i, p_j)\ ,\ J = \{j : w\left(|\boldsymbol{x}_j - \boldsymbol{x}_i|\right) \neq 0\} \tag{6.68}$$

This replacement improves the stability of the code by ensuring the interparticle repulsive force.[50]

6.3.3. *Boundary conditions*

In both SPH and MPS methods, solid boundaries such as walls or other fixed objects are represented by fixed particles, i.e., particles with zero velocity. The wall particles consist of two main layers: inner wall particles and a few lines of dummy particles as shown in Fig. 6.3. To solve the Poisson Pressure Equation for the inner wall particles, they are handled in a comparable manner as the wall particles in a grid-based method. To prevent the calculated particle number densities (in the MPS method) or particle densities (in the SPH method) at the inner wall boundary from becoming too small, dummy particles are introduced. Without the dummy particles, the inner wall particles would be recognized as a free surface.

With standard MPS or ISPH methods, the assessment of free-surface particles hinges on the simple fact that the calculated particle number density or particle density drops abruptly at the free surface. Thus, a single particle that satisfies the simple condition

⟨ MPS method ⟩

$$n_i < \beta n_0 \tag{6.69}$$

⟨ ISPH method ⟩

$$\rho_i < \beta \rho_0 \tag{6.70}$$

is considered a free-surface particle; for this particle, the zero pressure boundary condition is applied. In the above equations β is a constant chosen to be slightly below 1.0 (i.e., usually taken as 0.97 in the MPS method and 0.99 in the ISPH method).

Despite simplicity and ease of implementation, assessing the free surface simply based on the above mentioned criterion would not necessarily be accurate, especially in cases with violent fluid flows. With very turbulent flows, the calculated density or number density at some inner fluid particles may also drop below the specified threshold.

Khayyer *et al.*[42] proposed a simple and efficient criterion for a more accurate determination of free-surface particles. For a free-surface particle, a non-symmetric distribution of neighboring particles exist. For this reason, the summation of relative particle positions about either the x- or y-coordinates in the neighborhood of a target particle would be larger than the initial particle spacing or the diameter of one particle ($= d_0$). Accordingly, the accompanying criterion for free-surface assessment is introduced as

$$\left| \sum_{i \neq j} x_{ij} \right| > \alpha \quad \text{or} \quad \left| \sum_{i \neq j} y_{ij} \right| > \alpha \tag{6.71}$$

where $\alpha = d_0$. Equation (6.71) is applied together with Eq. (6.69) or (6.70) for an accurate and efficient assessment of free-surface particles. Alternative formulations to accurately assess the presence of a free surface in particle methods have been proposed by Lee *et al.*[52] and Ma and Zhou.[55]

6.3.4. *Neighboring particle search*

Both the SPH and MPS methods apply kernel functions with compact support, thus, a list of neighboring particles j for a target particle i must always be correct. An efficient and accurate particle-based calculation requires an efficient and suitable neighboring search strategy. The simplest and most direct search for neighboring particles is the so-called all-pair search algorithm. For each target particle i, the distance to another particle k is checked to see if particle k lies within the target particle's influential area to become a neighboring particle j for particle i. Despite its simplicity, the all-pair search algorithm is only applicable to small-scale problems, as it requires significant computational effort.

Most SPH-based calculations with a constant smoothing length appear to use the so-called linked list algorithm in which substantial savings in computational time are achieved by using cells as a bookkeeping tool. The same approach is commonly applied in both SPH and MPS calculations. For calculations consisting of adaptive smoothing lengths (or circles or spheres of influence with variable radii), another neighboring search strategy, namely, the tree search algorithm, has been found to be appropriate. In the tree search algorithm, ordered trees are created according to the particle positions. Once the tree structure is created, it can be applied efficiently to set up a list of neighboring particles.

As the most efficient and straightforward approach, the linked list algorithm will be briefly explained. A background Cartesian square-shaped grid is overlaid with the calculated domain. Then, the grid spacing is selected to be consistent with the compact support of the applied kernel. When using a B-Spline kernel, the size of each square cell (l_{bg}) is chosen as $2h$. In MPS-based simulations, the grid spacing is set according to the maximum radius of influence considered for different differential operator models, namely, the gradient and Laplacian operators. For target particle i, its neighboring particles j lie either in the same cell or in adjacent cells. Hence, the search is carried out over 9 cells (for a two-dimensional computation) to yield a candidate list of neighboring particles. In this manner, the list of neighboring particles is efficiently generated from the candidate list of neighboring particles.

The number of operations, N_{can}, to generate candidate list of neighboring particles would be[23]

$$N_{can} = aN + \frac{A_0}{l_{bg}^2} = aN + \frac{A_0}{(2h)^2} \tag{6.72}$$

where N is the number of particles, A_0 is the total area of the calculated domain and a is the number of operations required per particle. The first term of Eq. (6.72) denotes the number of required operations to identify the particle positions on the background grid, while the second term shows the required number of operations to initialize the screening grid. The required number of operations to search for neighboring particles in 9 cells is written as:

$$N_{nei} = bM_c N\,; \quad M_c = \frac{9l_{bg}^2}{\pi r_e^2}M = \frac{9}{\pi}M \tag{6.73}$$

where M is the number of neighboring particles in the circle of influence with radius r_e, M_c is the number of candidate particles in the neighboring particle search area, and b is the number of operations per particle. Accordingly, the total number of required operations for the neighboring particle search is written as[23]

$$N_{list} = N_{can} + N_{nei} = \left(a + \frac{9b}{\pi}M\right)N + \frac{A_0}{(2h)^2} \tag{6.74}$$

The order of operation count is approximately MN because a, b, and $A_0/(2h)^2$ are quite smaller than M, especially for problems with high resolution where a large number of particles is introduced. On the other hand, the number of operations with an all-pair search algorithm is written as

$$N_{list} \propto N^2 \tag{6.75}$$

Conclusively, the number of operations in a linked list algorithm would be quite less than that for an all-pair search, by a factor of M/N.

Koshizuka *et al.*[50] proposed a screening circle instead of a screening grid to generate the list of neighboring particles. In their study, the number of operations was proportional to $N^{1.5}$, resulting in a reduction of operations by a factor of $1/N^{0.5}$. Therefore, whenever $M < N^{0.5}$, employing a screening grid would be more efficient than employing a screening circle.

6.4. Sub-Particle Scale Turbulence Model

Sub-Grid Scale (SGS) turbulence models were first introduced in an Eulerian Large Eddy Simulation (LES) context (Chapter 2). Following the same concept, Gotoh *et al.*[30] introduced a Sub-Particle-Scale (SPS) turbulence model that is indispensable to particle-based simulations of turbulent flows.

A spatial filter is applied to express the velocity as a mean Particle Scale (PS) component, $\overline{u_l}$, and an SPS turbulent one, u_l'

$$u_l = \overline{u_l} + u_l' \tag{6.76}$$

Neglecting the Leonard term and the cross term, the PS momentum equation is written as:

$$\frac{D\overline{u_l}}{Dt} = -\frac{1}{\rho}\frac{\partial\overline{p}}{\partial x_l} + \nu\frac{\partial^2\overline{u_l}}{\partial x_m^2} - \frac{\partial(\overline{u_l' u_m'})}{\partial x_m} \tag{6.77}$$

Replacing the velocity and pressure with their PS components, and comparing Eq. (6.77) with the Navier–Stokes equation, the Reynolds stress term is found as an additional term. The Reynolds stress term, described by the eddy viscosity model, is expressed as follows:

$$-\overline{u_l' u_m'} = \nu_t\left(\frac{\partial\overline{u_l}}{\partial x_m} + \frac{\partial\overline{u_m}}{\partial x_l}\right) - \frac{2}{3}k\delta_{lm} \tag{6.78}$$

where ν_t is the kinematic eddy viscosity, k is the turbulent kinetic energy, and δ_{lm} is Kronecker's delta. Applying the Smagorinsky model[72] as a simplest form, the kinematic eddy viscosity and the energy dissipation rate are given in the following forms:

$$\nu_t = C_\nu k^{1/2}\Delta \tag{6.79}$$

$$\varepsilon = \frac{C_\varepsilon k^{3/2}}{\Delta} \tag{6.80}$$

where Δ is the filter width ($= d_0$) and C_ν, C_ε are constants. Assuming local equilibrium of the SPS turbulence,

$$-\overline{u_l' u_m'}\frac{\partial\overline{u_l}}{\partial x_m} = \nu_t\left(\frac{\partial\overline{u_l}}{\partial x_m} + \frac{\partial\overline{u_m}}{\partial x_l}\right)\frac{\partial\overline{u_l}}{\partial x_m} = \varepsilon \tag{6.81}$$

The kinematic eddy viscosity is derived from Eqs. (6.79) and (6.80) as

$$\nu_t = (C_s\Delta)^2\left\{\left(\frac{\partial\overline{u_l}}{\partial x_m} + \frac{\partial\overline{u_m}}{\partial x_l}\right)\frac{\partial\overline{u_l}}{\partial x_m}\right\}^{\frac{1}{2}} \tag{6.82}$$

where C_s is the Smagorinsky constant. The PS momentum equation, or Eq. (6.77), is closed by Eqs. (6.78) and (6.82). The constants in Eqs. (6.79), (6.80) and (6.82) must satisfy the following relation:

$$C_s^2 = C_\nu^{3/2}C_\varepsilon^{-1/2} \tag{6.83}$$

Despite popularity and applicability to a variety of turbulent flows, the Smagorinsky model has a few shortcomings: excessive dissipation, incorrect asymptotic behavior near solid surfaces, and adjustments of the Smagorinsky constant for laminar flow regions and flows with high shear.

The introduction of dynamic modeling[18] has propelled significant progress in the SGS modeling of non-equilibrium flows (see §2.3.6). In dynamic models, the coefficients of the model are a function of space and time, and they are dynamically determined during calculations. For instance, in the dynamic Smagorinsky model by Germano *et al.*,[18] two different filters, namely, the grid filter and the test filter, are employed to determine the model coefficient using the turbulence energy content on the smallest resolvable scale. The same concept can be applied to particle methods, resulting in dynamic SPS turbulence models. Prior to that, however, accurate and consistent numerical schemes are crucial for precise simulations with differential operator models.

Since the introduction of SPS turbulence models, particle methods provide potentially reliable and accurate simulations of violent, turbulent flows, including breaking waves, wave run-up, wave overtopping, wave impact on coastal structures, etc. (see Gotoh *et al.*;[22] Shao and Gotoh;[67] Dalrymple and Rogers;[13] Hori *et al.*[34]).

6.5. Accurate Particle Methods

Despite their robustness and wide-range of applicability, both SPH and MPS methods have several major drawbacks, mainly caused by the interpolation process of particle methods, i.e., local kernel-based interpolations on the basis of moving calculation points. Non-conservation of momentum, unphysical pressure fluctuations, and numerical instability are among the major drawbacks associated with particle methods. These drawbacks usually manifest themselves in forms of unphysical behaviors, such as unphysical fragmentations, clumpings and perturbations; sometimes they result in a complete blow up of the calculation. In this section, a concise review of the state-of-the-art of the improvements in particle methods is presented.

6.5.1. *Improvements with momentum conservation*

All fluid flow computations are based on the fundamental principles of physics, including conservation of mass and momentum. Particle methods are not an exception; however, due to the particle-based discretization, local (and hence global) conservation of momentum may not be guaranteed with a particle method, unless special attention is given to interparticle forces.

Momentum conservation has been an important theme in SPH research (e.g., Monaghan;[59] Bonet and Lok[6]). Monaghan[59,62] proposed modified

SPH formulations to give interparticle interacting forces in a momentum-conservative form. Bonet and Lok[6] set forth a discrete variational SPH approach that ensures the balance of both linear and angular momentum. They proposed corrective techniques to ensure the invariance of potential energy with respect to rigid body motions, and thus, to guarantee the conservation of both linear and angular momentum. Khayyer *et al.*[38] applied the same variational approach to ensure the conservation of angular momentum in ISPH-based simulations.

As highlighted by Bonet and Lok,[6] Khayyer *et al.*,[41] and Khayyer and Gotoh,[38] the key issue for conservation of linear momentum in a particle-based simulation is that the internal interparticle forces must be anti-symmetric, or equal in magnitude, and opposite in direction. For exact conservation of angular momentum, internal interparticle forces must be radial, or co-linear, with the position vector, in addition to being anti-symmetric. In the ISPH method of Shao and Lo,[69] both pressure and viscous interacting forces are anti-symmetric. Hence, exact conservation of linear momentum is guaranteed. However, the viscous interacting forces are not radial due to the anisotropic nature of viscous stresses; conservation of angular momentum is not ensured. With the MPS method, the situation is even worse as the pressure-based interparticle forces are not anti-symmetric. Accordingly, neither linear nor angular momentum is preserved in MPS-based simulations.

By applying the same variational approach introduced by Bonet and Lok,[6] Khayyer *et al.*[41] derived a corrective matrix for the viscous stress model in ISPH-based simulations. Introducing corrective terms guarantees the invariance of potential energy with respect to rigid body motions and thus ensures the conservation of angular momentum. Further, linear velocity fields are exactly calculated when such a variational formulation is being employed. The modified ISPH method was named Corrected Incompressible SPH (CISPH) analogous to the Corrected SPH (CSPH) method proposed by Bonet and Lok.[6]

By considering an auxiliary point k on the midpoint of the position vector of particle i and its neighboring particle j, Khayyer and Gotoh[38] derived an anti-symmetric and radial pressure gradient model in the MPS framework. This novel formulation, called Corrected MPS (CMPS) method, is expressed as

$$\nabla p|_i = \frac{D_s}{n_0} \sum_{j \neq i} \frac{(p_i + p_j) - (\hat{p}_i + \hat{p}_j)}{|\boldsymbol{x}_j - \boldsymbol{x}_i|^2} (\boldsymbol{x}_j - \boldsymbol{x}_i) w (|\boldsymbol{x}_j - \boldsymbol{x}_i|) \tag{6.84}$$

The above equation was validated by a set of numerical simulations, including the evolution of an elliptical water droplet, and plunging wave breaking and resultant splash-up corresponding to the experiment by Li and Raichlen.[53] To further improve results, Khayyer and Gotoh[38] applied a tensor-type, strain-based viscosity term in place of the original viscosity model of the MPS method. The abbreviation CMPS-SBV means CMPS with Strain-Based Viscosity.

Figure 6.4 illustrates the snapshots from standard MPS, CMPS and CMPS-SBV simulations, as compared with their corresponding laboratory photographs.[53] The figure shows the enhanced performance of the CMPS method with respect to the original MPS method. As a result of exact conservation of linear momentum and improved preservation of angular momentum, the snapshots for the CMPS portray a clearer image of the plunging jet and the air chamber beneath it with less unphysical particle scattering as seen in the snapshots for the MPS. Further, the CMPS provides a refined reproduction of how splash-up forms and develops as compared with the MPS method. For instance, the angle of the reflected jet and the gap between the incident and reflected jets agree better with those in the experiment. As splash-up progresses, the reflected jet curls back towards the incident jet and eventually becomes almost vertical. Although this fact is difficult to be reproduced with the CMPS method, applying a strain-based viscosity using the CMPS-SBV method has reproduced the entire stages of splash-up formation, development, and backward curling. In particular, the backward curling of the splash-up is well simulated by the CMPS-SBV. The reflected jet angle, the gap between the plunging jet and the reflected jet, as well as the geometrical shape of the air chamber are all in good agreement with the experimental observations.

6.5.2. *Control of pressure fluctuations*

One of the major drawbacks associated with particle methods is the presence of unphysical fluctuations in the pressure field. This problem has been addressed by some researchers.[1,10,23,39,43,45,58] Colagrossi and Landrini[10] presented a relatively improved pressure calculation using a modified WC-SPH method.

Shibata and Koshizuka[70] carried out a three-dimensional MPS-based calculation of water impacting ship hulls. In their study, an acceptable simulation-experiment agreement was achieved in terms of water surface elevation. Nevertheless, even after spatial averaging, the peak impact pres-

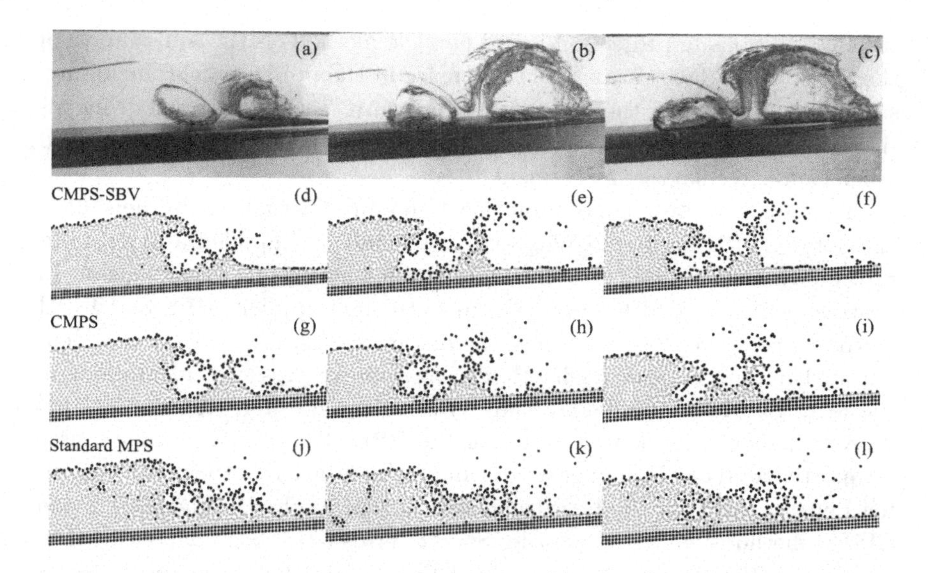

Fig. 6.4. Plunging breaker simulated by standard MPS, CMPS and CMPS-SBV.

sure was underestimated by about 50%, and an acceptable pressure value could not be obtained. Therefore, the original MPS method does not seem to be a reliable tool to calculate pressure associated with violent wave impacts, unless modifications and corrections are made.

As one of the earliest efforts to minimize unphysical pressure fluctuations in the MPS method, Koshizuka *et al.*[49] considered a modified source term of the PPE as follows

$$\nabla^2 p_{k+1} = \frac{\rho_0}{\Delta t^2}\left(\frac{n_k^* - n_0}{n_0} - \frac{1}{\rho_0 C_s^2} p_{k+1}\right) \tag{6.85}$$

where the second term on the right hand side represents the effect of compressibility. In fact, it was shown that a slight level of compressibility would help decrease in unphysical pressure fluctuations. However, allowing for compressibility would also result in an unphysical damping of the flow energy as well as a clear violation of volume conservation, especially in long-term simulations. Besides, the main source of pressure fluctuations stems from applying simplified numerical schemes and assumptions made during derivation.

As shown by Khayyer and Gotoh,[39] Khayyer *et al.*,[42] Kondo and Koshizuka,[45] and Khayyer and Gotoh,[43] the main cause of unphysical pressure fluctuations in projection-based particle methods is the source term of the PPE. Other factors contribute to noise in the pressure field, including: simplified schemes for the Laplacian of pressure,[40] non-conserved or inconsistent (incomplete) schemes for the pressure gradient,[41,43] and simplified schemes for free-surface assessment,[42] etc.

To resolve the problem of unphysical pressure fluctuations in their fully-explicit WCSPH method, Colagrossi and Landrini[10] applied a more accurate interpolation scheme to calculate the instantaneous density field. In contrast to the WCSPH approach, in both semi-implicit MPS and ISPH methods, pressure is a function of the instantaneous time variation of density rather than density itself. Hence, to improve pressure calculations in the MPS and ISPH methods, Khayyer and Gotoh[39] and Khayyer *et al.*[42] derived higher order source terms for the PPE. These higher order source terms are based on the higher order time differentiation of the particle number density (MPS method) or the density (SPH method). The CMPS and CISPH methods with Higher order Source terms (-HS) were named CMPS-HS and CISPH-HS, respectively. The PPEs with higher order source terms for the MPS and ISPH methods are formulated by Eqs. (6.86) and (6.87), respectively.

⟨MPS method⟩

$$\nabla^2 p_{k+1}\big|_i = -\frac{\rho_0}{n_0 \Delta t} \sum_{i \neq j} \frac{r_e}{r_{ij}^3} (\boldsymbol{x}_{ij} \cdot \boldsymbol{u}_{ij})^* \tag{6.86}$$

⟨ISPH method⟩

$$\nabla^2 p_{k+1}\big|_i = \frac{\rho^*}{\rho_0 \Delta t} \sum_{i \neq j} (m_j \nabla_i W_{ij} \cdot \boldsymbol{u}_{ij})^* \tag{6.87}$$

The superscript * in Eqs. (6.86) and (6.87) denotes the fact that all the physical quantities of the source term are calculated at the prediction step.

Figure 6.5 shows the step-by-step improvements in the calculation of hydrostatic pressure by considering weak compressibility, higher order source terms, and a corrected pressure gradient model. The step-by-step enhancements in simulating a flip-through process are illustrated in Fig. 6.6. In addition to the pressure field, the water surface profile extracted from experimental photographs[33] is also presented in each snapshot. Analogous to the hydrostatic pressure simulations, all of the modified MPS methods have improved pressure calculations compared to the standard MPS method.

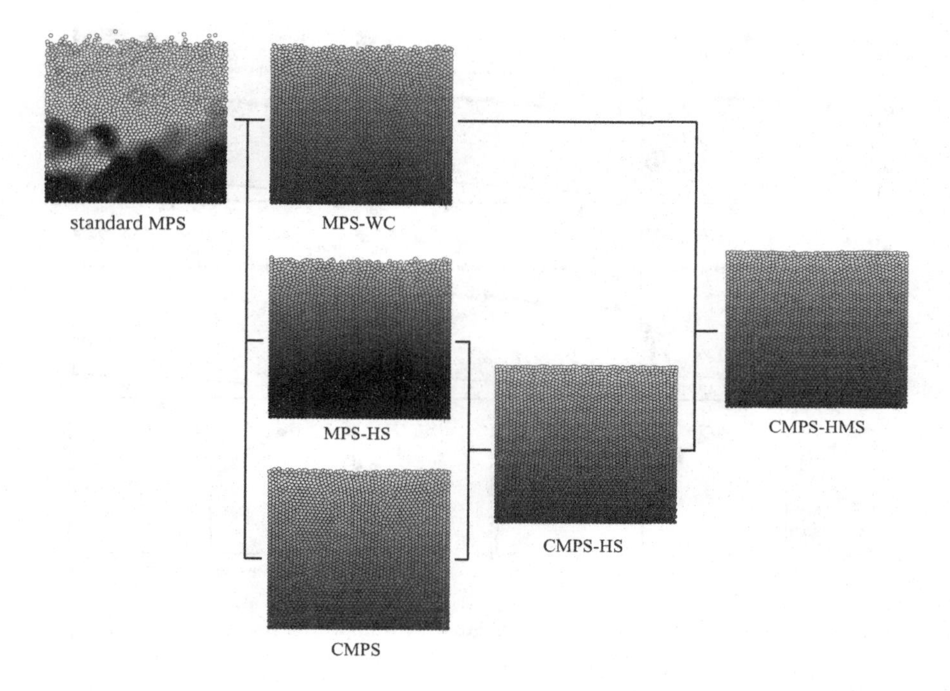

Fig. 6.5. Hydrostatic pressure calculated with improved particle methods.

When a momentum-conserved pressure gradient term is used with the CMPS method, relative particle positions are controlled; this leads to a more-accurate and less-fluctuating particle number density field, resulting in more of a physical pressure field. The pressure field by the MPS-HS method tends to be more smoothly distributed than that by the CMPS method. Meanwhile, in an MPS-HS snapshot, a few inner zero-pressure particles are evident. A combination of modifications to the CMPS and MPS-HS methods has resulted in a more enhanced spatial distribution of pressure as well as a water surface profile that agrees better with experimental results.

The CMPS-HS snapshot is characterized by distinctive pressure contours quite similar to those computed by Cooker and Peregrine.[11] The result by the MPS method with Weak Compressibility (MPS-WC) is superior to that by standard MPS. The snapshot of the Corrected MPS with a Higher order Modified Source term (CMPS-HMS), which introduced a slight level of compressibility into the CMPS-HS, appears to be similar to

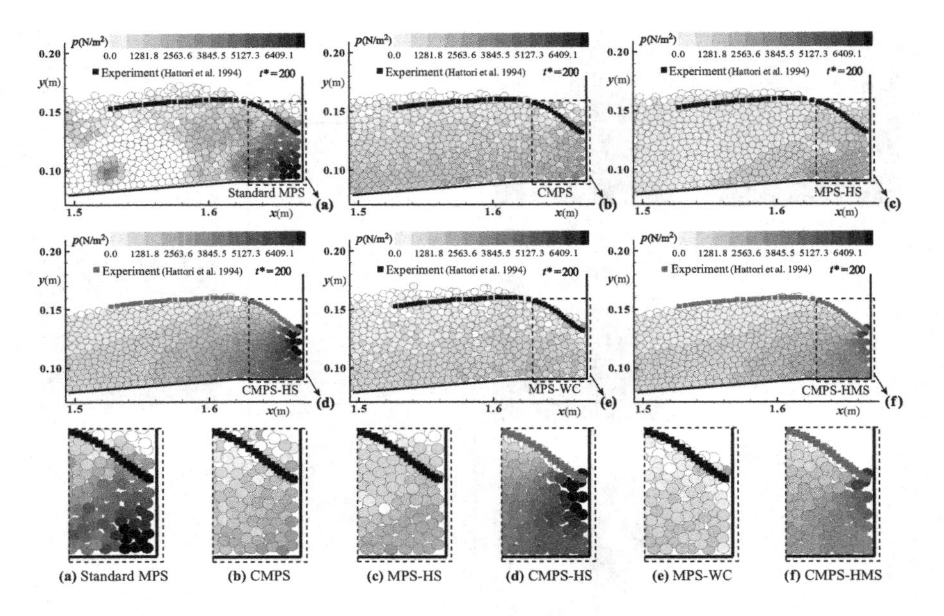

Fig. 6.6. Flip-through impact simulated with improved particle methods.

the CMPS-HS snapshot, especially in terms of the free surface profile. However, the pressure field seems to be better reproduced by the CMPS-HMS, as this method has resulted in a more localized maximum pressure close to the still water level, which is in qualitative agreement with the computation by Cooker and Peregrine.[11]

The effectiveness of the above mentioned modifications has also been verified quantitatively by comparing the time histories of calculated pressure with their corresponding experimental ones.[39] Despite showing significant improvements with respect to the original MPS method, both CMPS-HS and CMPS-HMS methods appear to provide only approximate estimations of the wave impact pressure; the presence of numerical noise was still obvious. This fact indicated the need for further improvements.

Another step towards improving pressure calculations using a projection-based particle method is to apply a more accurate Laplacian model to discretize the Laplacian of pressure in the PPE. Khayyer and Gotoh[40] highlighted the importance of the mathematical consistency of the Laplacian model and discretized source term of the PPE, and they derived a Higher order Laplacian model (-HL) for the MPS method. The higher order Laplacian model was derived by meticulously taking the divergence of an SPH gradient model:[59]

$$\nabla \cdot \langle \nabla \phi \rangle_i = \frac{1}{n_0} \sum_{i \neq j} \left(\nabla \phi_{ij} \cdot \nabla w_{ij} + \phi_{ij} \nabla^2 w_{ij} \right) \tag{6.88}$$

where

$$\langle \nabla \phi \rangle_i = \frac{1}{n_0} \sum_{i \neq j} (\phi_j - \phi_i) \nabla w_{ij} = \frac{1}{n_0} \sum_{i \neq j} \phi_{ij} \, \nabla w_{ij} \tag{6.89}$$

In two-dimensions, the higher order Laplacian is formulated as

$$\nabla \cdot \langle \nabla \phi \rangle_i = \frac{1}{n_0} \sum_{i \neq j} \left(\phi_{ij} \frac{\partial^2 w_{ij}}{\partial r_{ij}} - \frac{\phi_{ij}}{r_{ij}} \frac{\partial w_{ij}}{\partial r_{ij}^2} \right) \tag{6.90}$$

By considering the standard MPS kernel, Eq. (6.90) is simplified to

$$\nabla \cdot \langle \nabla \phi \rangle_i = \frac{1}{n_0} \sum_{i \neq j} \left(\frac{3 \, \phi_{ij} r_e}{r_{ij}^3} \right) \tag{6.91}$$

The enhanced performance of the higher order Laplacian model was verified with simulations of designed exponentially excited sinusoidal pressure oscillations,[40,43] as well as violent sloshing flows corresponding to the experiment by Kishev *et al.*[44]

In an attempt to suppress the unphysical pressure fluctuations in the MPS method, Kondo and Koshizuka[45] suggested a multi-term source for the PPE. The objective of the multi-term source was to obtain smooth time variations of the particle number density field (n-field) while keeping it invariant. Despite resulting in substantial improvements, the numerical scheme proposed by Kondo and Koshizuka[45] had several shortcomings: presence of unknown coefficients that require appropriate calibration, application of accurate first-order schemes for discretization, and ignorance of the free surface.

Khayyer and Gotoh[43] proposed a distinct modified source term for the PPE to enhance and stabilize pressure calculations. To derive the modified source term, they revisited the derivation of the PPE in the MPS method. The modification of the PPE was comprised of a high-order main term equivalent to that in the CMPS-HS[39] and two high-order error-compensating terms with dynamic coefficients as functions of the instantaneous flow field. Further, the effect of the free surface was carefully considered while deriving the newly proposed scheme; the derivation is explained here.

The MPS method is founded on a Helmholtz–Hodge decomposition of a vector field where an intermediate velocity field, $\boldsymbol{u}^*$, is considered to be composed of a divergence free velocity field and the gradient of a scalar field (Fig. 6.7), i.e.,

$$\boldsymbol{u}^* = \boldsymbol{u}_{k+1} + \frac{\Delta t}{\rho}\boldsymbol{\nabla} p \tag{6.92}$$

The above vector decomposition brings about a two-step prediction-correction solution process. As it was discussed before, in the prediction step, an intermediate velocity field $\boldsymbol{u}^*$ is obtained explicitly by considering viscous and gravitational forces. At this step, the incompressibility of the fluid is violated, i.e., the divergence of $\boldsymbol{u}^*$ would not be equal to zero. Hence, a new corrective velocity field, $\boldsymbol{u}^{**}$, is computed to project $\boldsymbol{u}^*$ onto a divergence free space so that the final velocity field, $\boldsymbol{u}_{k+1}$, would be divergence free. The new corrective velocity field is computed by considering the temporal time rate of change of the particle number density on the basis of the continuity equation

$$\frac{1}{n_0}\left(\frac{Dn}{Dt}\right)^* + \boldsymbol{\nabla}\cdot\boldsymbol{u}^{**} = 0 \tag{6.93}$$

From Eqs. (6.92) and (6.93),

$$\boldsymbol{\nabla}\cdot\boldsymbol{u}^{**} = \boldsymbol{\nabla}\cdot(\boldsymbol{u}_{k+1} - \boldsymbol{u}^*) = \boldsymbol{\nabla}\cdot\left(-\frac{\Delta t}{\rho}\boldsymbol{\nabla} p\right) = -\frac{1}{n_0}\left(\frac{Dn}{Dt}\right)^* \tag{6.94}$$

Accordingly, the correction step would consist of solving a PPE to obtain the updated pressure and corrective velocity fields. From Eq. (6.94), the PPE is obtained as

$$\frac{\Delta t}{\rho}\left(\boldsymbol{\nabla}^2 p_{k+1}\right)_i = \frac{1}{n_0}\left(\frac{Dn}{Dt}\right)^*_i \tag{6.95}$$

From Eqs. (6.94) and (6.95), it is clear that the accuracy of the numerical solutions would depend on several items, including: (i) Accuracy of the source term of the PPE (in other words, how accurate is the time rate of volume change at the end of a prediction step, i.e., $1/n_0\,(Dn/Dt)^*$; (ii) Accuracy/consistency of the numerical schemes to discretize the differential operators (such as the Laplacian and gradient); (iii) Accuracy of the numerical schemes to integrate with respect to time; and (iv) Accuracy of the solver of the simultaneous linear equations obtained after discretization of Eq. (6.95).

As a result of the numerical errors arising from incomplete, inconsistent or non-conservative particle-based numerical schemes, the approximated

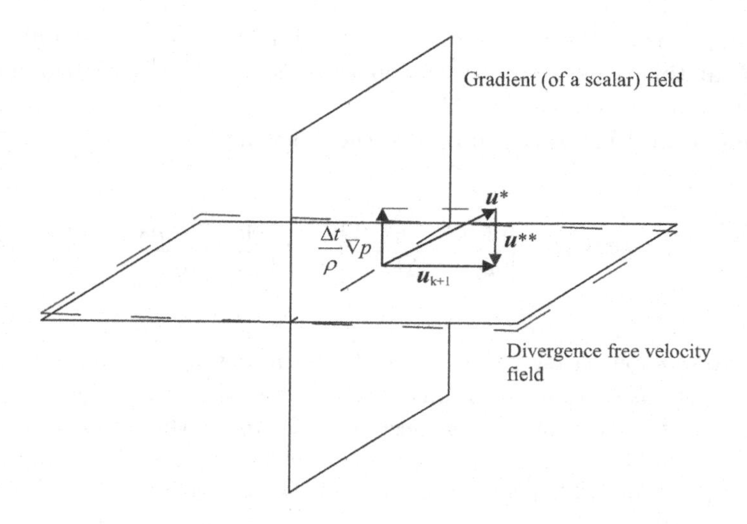

Fig. 6.7. Schematic expression of the projection method.

velocity field $\boldsymbol{u}_{k+1}$, would not be divergence free and fluid incompressibility (at time step $k+1$) would not be perfectly satisfied. In other words, at each time step, the calculated velocity field $\boldsymbol{u}_{k+1}$ slightly deviates from the divergence free field, as shown schematically by a dashed plane in Fig. 6.7. Further, as time proceeds, this instantaneous deviation would bring about an accumulative density error.

To obtain instantaneous, divergence-free velocity fields, overall volume conservation and accurate/stabilized pressure/velocity fields, Hu and Adams[35] and Kondo and Koshizuka[45] introduced Error-Compensating Source (ECS) terms for the PPE. The ECS terms should be measures for both instantaneous and accumulative violations of fluid incompressibility. Thus, the modified PPE is introduced as

$$\begin{aligned}\frac{\Delta t}{\rho}\left(\boldsymbol{\nabla}^2 p_{k+1}\right)_i &= \frac{1}{n_0}\left(\frac{Dn}{Dt}\right)_i^* + ECS \\ &= \frac{1}{n_0}\left(\frac{Dn}{Dt}\right)_i^* + \text{func}\left[\frac{1}{n_0}\left(\frac{Dn}{Dt}\right)_i^k, \frac{1}{\Delta t}\frac{n^k - n_0}{n_0}\right]\end{aligned} \tag{6.96}$$

The first error-compensating parameter in Eq. (6.96) is related to the instantaneous time variation of the particle number density at time step k or the divergence of the velocity at this time step (both of which should be zero theoretically). The second parameter reflects the deviation of n at time step k from the constant n_0, or the time rate of overall volumetric change at time step k, i.e., it accounts for the accumulative error in the particle number density.

Kondo and Koshizuka[45] proposed the following ECS term for the PPE.

$$ECS = \alpha \left[\frac{1}{n_0} \frac{n^k - n^{k-1}}{\Delta t} \right] + \beta \left[\frac{1}{\Delta t} \frac{n^k - n_0}{n_0} \right] \tag{6.97}$$

The error-compensating terms in Eq. (6.97) represent first-order-accurate approximations of time variations of n and the time rate of overall volumetric change at time step k, respectively. There are three major issues to consider with Eq. (6.97). First, as pointed out before, the error compensating parts are multiplied by unknown coefficients. Kondo and Koshizuka[45] calibrated the coefficients and proposed some typical coefficients by performing a hydrostatic pressure calculation. Nevertheless, since generated errors in the source term of the PPE depend on instantaneous flow fields, instantaneous (relative) particle positions, and their spatial/time variations, the calibrated coefficients obtained from a hydrostatic pressure calculation would not necessarily be appropriate for other simulation cases, particularly violent fluid flows. Second, the numerical schemes applied to approximate the time derivatives of n are first-order-accurate ones. This may introduce further numerical errors, especially in simulations of violent fluid flows. Third, the effect of the free-surface has not been considered in the derivation of Eq. (6.97).

Khayyer and Gotoh[43] also considered a linear summation form of the error-compensating parameters, i.e.,

$$\left. \begin{aligned} ECS &= \alpha \left[\frac{1}{n_0} \left(\frac{Dn}{Dt} \right)_i^k \right] + \beta \left[\frac{1}{\Delta t} \frac{n^k - n_0}{n_0} \right] \\ \alpha &= \left| \frac{n^k - n_0}{n_0} \right| \quad ; \quad \beta = \left| \frac{\Delta t}{n_0} \left(\frac{Dn}{Dt} \right)_i^k \right| \end{aligned} \right\} \tag{6.98}$$

in which the coefficients α and β were formulated on the basis of instantaneous flow features. Moreover, the main part, as well as the error compensating parts, of the multi-term PPE were discretized using high-order

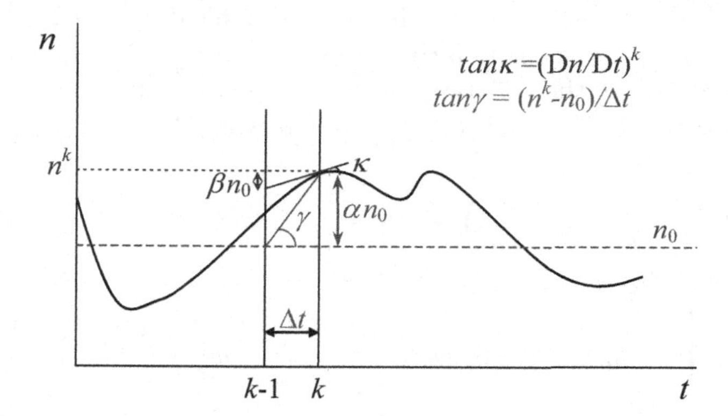

Fig. 6.8. Schematic figure of particle number density variations.

numerical schemes. Equation (6.98) results in

$$\frac{\Delta t}{\rho}\left(\nabla^2 p_{k+1}\right)_i = \begin{cases} \dfrac{1}{n_0}\left(\dfrac{Dn}{Dt}\right)_i^* \pm \dfrac{n^k - n_0}{n_0}\left[\dfrac{2}{n_0}\left(\dfrac{Dn}{Dt}\right)_i^k\right] \\ \qquad\qquad \text{if } \left(\dfrac{Dn}{Dt}\right)^k \cdot \dfrac{n^k - n_0}{n_0} > 0 \\ \dfrac{1}{n_0}\left(\dfrac{Dn}{Dt}\right)_i^* \\ \qquad\qquad \text{if } \left(\dfrac{Dn}{Dt}\right)^k \cdot \dfrac{n^k - n_0}{n_0} < 0 \end{cases} \tag{6.99}$$

There are two indices, namely the instantaneous time variation of particle number density, $(Dn/Dt)^k$, and the deviation of particle number density from the constant n_0, $(n^k - n_0)/n_0$. When the two indices have the same sign $(+,+)$ or $(-,-)$, the ECS term works. The ECS term vanishes when the two indices have different signs $(+,-)$ or $(-,+)$.

Focusing on Eq. (6.99) and Fig. 6.8, it is clear that we are dealing with the velocity of particle number density variations as well as its deviations (distance) from the initial constant one (n_0). In Eq. (6.99), the absolute values of coefficients denote the intensity for variations in the error-compensating terms, and the negative or positive signs are determined by the error-compensating terms themselves, depending on the instantaneous state of the volume change and the instantaneous time rate of compression/expansion. It should be added that one advantage of Eq. (6.99) with respect to Eq. (6.97) would be that the right hand side of Eq. (6.99) is independent of the calculation time step Δt and thus, this equation would be less sensitive to the variations in Δt and the time integration process.

Khayyer and Gotoh[43] proposed the following PPE considering a free surface.

$$\frac{\Delta t}{\rho}\left(\nabla^2 p_{k+1}\right)_i = \frac{1}{n_0}\left(\frac{Dn}{Dt}\right)_i^* + ECS$$

$$ECS = \begin{cases} \left|\dfrac{n^k - n_0}{n_0}\right| \left[\dfrac{1}{n_0}\left(\dfrac{Dn}{Dt}\right)_i^k\right] + \left|\dfrac{\Delta t}{n_0}\left(\dfrac{Dn}{Dt}\right)_i^k\right| \left[\dfrac{1}{\Delta t}\dfrac{n^k - n_0}{n_0}\right] & \text{if } n^k > \gamma n_0 \\ \left|\dfrac{n^k - (n_i)_S}{(n_i)_S}\right| \left[\dfrac{1}{n_0}\left(\dfrac{Dn}{Dt}\right)_i^k\right] + \left|\dfrac{\Delta t}{n_0}\left(\dfrac{Dn}{Dt}\right)_i^k\right| \left[\dfrac{1}{\Delta t}\dfrac{n^k - (n_i)_S}{(n_i)_S}\right] & \text{if } n^k \leq \gamma n_0 \end{cases} \tag{6.100}$$

where γ $(= 0.90)$ is a constant smaller than that in the free-surface criterion. In Eq. (6.100), $(n_i)_S$ corresponds to an initial particle number density of a target particle i that is initially located at or close to the free-surface. At the beginning of a calculation, n_i should be calculated for all the particles in the calculation domain. Certainly, for particles at and close to the free-surface, the initially calculated n_i values would be much smaller than n_0, which is calculated from the initial arrangement of equally-spaced inner particles. In the case of violent fluid flows, one particle close to the free-surface may not remain close to that region during the calculation. Thus, $(n_i)_S$ should be updated every N (e.g., 20) time steps. Khayyer and Gotoh[43] highlighted that their approach is not solely limited to the MPS context, and an equivalent form of Eq. (6.100) can be derived for other projection-based particle methods including the ISPH method.[52,69]

6.5.3. *Other improvements*

Khayyer and Gotoh[43] showed that MPS schemes are vulnerable to unphysical behaviour in a tensile regime or in the presence of changes in internal stresses. This instability appears to be similar to the so-called "tensile instability" in the SPH method that has been widely studied by many researchers.[5,8,32,57,61,71,74] Bonet and Kulasegaram[5] demonstrated that tensile instability is a property of the elastic fluid whereby the stress tensor is isotropic and the pressure is a function of density (or more precisely the volume ratio).

To resolve the problem of tensile instability in the SPH method, Monaghan[61] and Gray *et al.*[32] proposed artificial repulsive forces proportional to the fluid pressure and the stress tensor, respectively. By applying a similar approach, Kondo *et al.*[46] improved the stability of a Hamiltonian particle method to simulate elastic structures. Dilts[14] showed that accurately estimating derivatives is a key point in removing tensile instability. Belytschko *et al.*[2] showed that carefully implementing stress points removes zero-energy modes, but the implementation does not generally eliminate tensile instability. Belytschko and Xiao[3] highlighted the fact that tensile instability occurs when an Eulerian kernel is used with a Lagrangian description of motion. They showed that this instability is eliminated when the kernel is a function of material coordinates (i.e., a Lagrangian kernel). It was also found that the best approach to stabilize particle-based methods is to use Lagrangian kernels with stress points.

However, apart from the increased complexity of mathematical formulations, when using kernels with stress points, the stability and convergence depends on the distribution of particles in the domain; a poor convergence rate would be obtained for irregular particle distributions.[17] The problem associated with Lagrangian kernels is that they may not tolerate large deformations that may occur in fluid flows,[3] particularly, in violent flows.

Khayyer and Gotoh[43] showed that tensile instability can be efficiently minimized in an MPS-based simulation provided that the pressure and its gradient are accurately calculated. Therefore, the first key step to eliminate tensile instability is related to suppressing unphysical numerical fluctuations by applying consistent, accurate numerical schemes to discretize the source term of the PPE and the Laplacian of pressure, as well as incorporating appropriate ECS terms and accurate time integration schemes.

The second major step to eliminate tensile instability is to derive accurate numerical schemes to approximate the pressure gradient. An accurate calculation of the pressure gradient leads to more accurate descriptions of particle motion and minimizes particle disorder as well as perturbations in particle motion. A more precise approximation of the pressure gradient can be achieved by focusing on the Taylor series expansion of pressure at a neighbouring particle j with respect to the pressure at target particle i (as shown by Chen *et al.*;[8] Oger *et al.*[65]). Incorporating this expansion,

Khayyer and Gotoh[43] derived a corrective matrix formulated as

$$\boldsymbol{C}_{ij} = \begin{pmatrix} \sum V_{ij} \frac{w_{ij} x_{ij}^2}{r_{ij}^2} & \sum V_{ij} \frac{w_{ij} x_{ij} y_{ij}}{r_{ij}^2} \\ \sum V_{ij} \frac{w_{ij} x_{ij} y_{ij}}{r_{ij}^2} & \sum V_{ij} \frac{w_{ij} y_{ij}^2}{r_{ij}^2} \end{pmatrix}^{-1} \tag{6.101}$$

where

$$V_{ij} = \frac{1}{\sum_{i \neq j} w_{ij}} \tag{6.102}$$

Accordingly, the corrected gradient model of the original MPS method was expressed as

$$\langle \boldsymbol{\nabla} p \rangle_i = \frac{D_s}{n_0} \sum_{j \neq i} \frac{p_j - p_i}{|\boldsymbol{r}_j - \boldsymbol{r}_i|^2} (\boldsymbol{r}_j - \boldsymbol{r}_i) \boldsymbol{C}_{ij} w\left(|\boldsymbol{r}_j - \boldsymbol{r}_i|\right) \tag{6.103}$$

A shortcoming of the above equation is related to the non-exact conservation of momentum, since the resulting interparticle pressure forces are not anti-symmetric.[38,41] Nevertheless, as pointed out by Oger *et al.*,[65] the numerical errors arising from a non-exact conservation of momentum might be quite negligible compared to those from approximating the pressure gradient, particularly in the presence of negative pressure fields or changes in stress states.

The enhanced performance of Eq. (6.103), namely the Gradient Correction (GC), were shown with a number of numerical tests, including a square patch of fluid subjected to purely rigid rotation,[9] a jet impinging on a flat plate[58] and a violently sloshing flow (Kishev *et al.*[44]).

Figure 6.9 depicts a series of snapshots by the MPS and modified MPS methods corresponding to the evolution of a square patch of fluid subjected to a rigid rotation. Figure 6.9(a) shows a typical snapshot by the MPS method at $t = 0.064$ s characterized by a spurious pressure field and dispersed particles at the patch corners. For this specific test, the MPS simulation was reported to break up fully at $t = 0.084$ s.

Application of a higher order source term with the MPS-HS method improved the stability of the simulation as it proceeded up to $t = 0.146$ s (Fig. 6.9(b)). From Fig. 6.9(c), employing ECS terms in the PPE further improved the stability of the MPS-HS method; however, the MPS-HS-ECS simulation failed to predict the evolution of the patch until $t = 0.20$ s. On the other hand, the MPS-HS-GC method, which benefited from a corrected pressure gradient term, provided a more accurate reproduction of the fluid

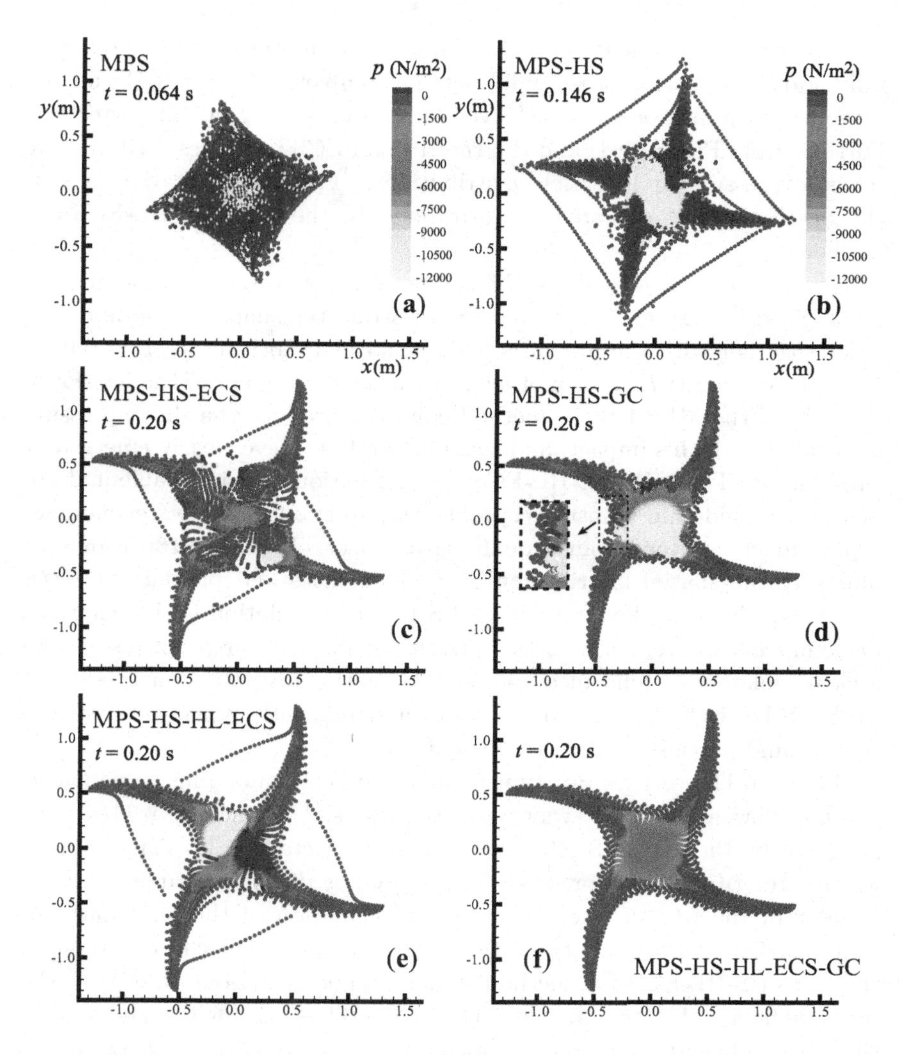

Fig. 6.9. Evolution of a square patch of fluid simulated with modified MPS methods.

patch at $t = 0.20$ s (Fig. 6.9(d)). However, at this particular instant the reproduced pressure field with the MPS-HS-GC simulation was fairly unphysical and the boundaries near the center of the patch were characterized by dispersed particles. Comparing Fig. 6.9(c) with (e), applying a higher order Laplacian produced a snapshot in better qualitative agreement with those by Colagrossi[9] and Fang *et al.*[16]

Nevertheless, the MPS-HS-HL-ECS simulation still failed to provide a fully stable simulation of the fluid patch. Comparing Figs. 6.9(e) and (f), a significant improvement is achieved by applying the gradient correction. Further, from Figs. 6.9(d) and (f), the HL and ECS terms resulted in more realistic pressure and particle distributions. From a qualitative aspect, the best snapshot appeared to correspond to the MPS-HS-HL-ECS-GC method.

Figure 6.10 corresponds to the simulation of a water jet impinging on a flat plate.[58] Figures 6.10(a) and (b) illustrate the snapshots of fluid particles together with the pressure fields at normalized time $tU/L = 0.9504$ (U: velocity of jet; L: length of jet), just after the impact of the jet. From this figure, the MPS-HS-HL method cannot reproduce the shock pressure associated with this impact, and some instabilities have clearly emerged at this instant. The MPS-HS-HL-ECS-GC simulation remains stable and appears to provide an almost acceptable reproduction of the water-hammer type impact, at least from a qualitative aspect. Figures 6.10(c) and (d) illustrate the spatial distribution of fluid particles and pressure at $tU/L = 1.3464$. The snapshot from the MPS-HS-HL simulation is characterized by significant discrepancies in the pressure field. This snapshot also shows a clear violation of fluid incompressibility at this instant. Conversely, the MPS-HS-HL-ECS-GC method provides a significantly enhanced distribution of fluid particles and pressure field.

Figure 6.11 portrays two typical snapshots corresponding to a violent, sloshing flow induced by sway-type excitations.[44] From Fig. 6.11(a), the snapshot by the MPS-HS-HL simulation is characterized by some unphysical irregularities in the pressure field as well as dispersed particles in the vicinity of the free-surface. The dispersive motion of the fluid particles is intensified in the vicinity of the impact region. On the other hand, the MPS-HS-HL-ECS-GC method shows a refined distribution of both the pressure field and water particles. This method has provided a one-particle-thick layer of the free-surface and has resulted in a qualitatively acceptable reproduction of pressure at the impact region.

6.6. Applications

6.6.1. *Wave breaking and overtopping*

Particle methods are Lagrangian methods that do no necessarily require a grid to solve; as such, particle methods are well suited to simulate hy-

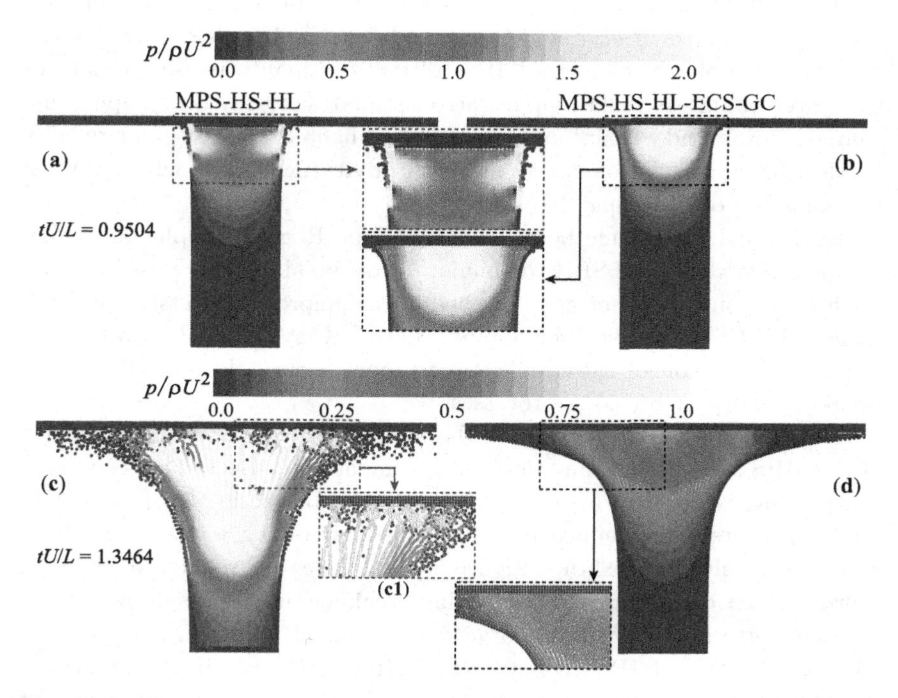

Fig. 6.10. Water jet impinging on a flat platesimulated by MPS-HS-HL-ECS-GC method.

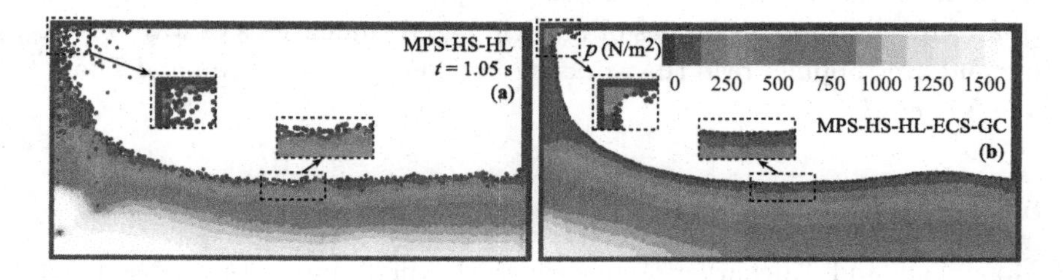

Fig. 6.11. Violent sloshing flow simulated by MPS-HS-HL-ECS-GC method.

drodynamic flow with large deformation and fragmentation. Some typical examples of these types of flow include wave breaking and post-breaking in the surf zone, as well as wave overtopping of coastal and offshore structures.

The first works to conduct particle-based wave breaking simulations are those by Koshizuka *et al.*[50] and Gotoh and Sakai.[25] Gotoh *et al.*[23] demonstrated the applicability of the MPS method for qualitative simulations of wave overtopping as well as quantitative estimates of wave overtopping discharge. Gotoh and Sakai[26] and Gotoh[20] highlighted key issues for reliable particle-based simulations of violent hydrodynamic flows, including wave breaking and overtopping.

In the field of SPH methods, Dalrymple and Rogers[13] applied a weakly compressible version of SPH to simulate wave breaking and post-breaking in the surf zone. Shao *et al.*[68] applied an incompressible version of SPH, namely ISPH,[69] to wave overtopping. Later, Khayyer *et al.*[41] presented a corrected version of ISPH, CISPH, to more accurately reproduce wave breaking and post-breaking processes.

Khayyer and Gotoh[38] revisited the momentum-conservation properties of the MPS formulations and derived a momentum-conserved pressure gradient model referred to as a CMPS, mentioned before in §6.5.1. Khayyer and Gotoh presented refined reproductions of plunging waves at breaking and their resultant splash-up. Figure 6.4 illustrates the refined MPS-based reproductions of a plunging wave during breaking and the resultant splash-up that corresponds to the laboratory experiment by Li and Raichlen.[53] The abbreviation CMPS-SBV denotes CMPS with a Strain-Based Viscosity. Gotoh *et al.*[24] carried out refined reproductions in three-dimensions of plunging wave breaking and resultant splash-up using a three-dimensional CMPS method. Figure 6.12 illustrates typical snapshots corresponding to three-dimensional CMPS simulations of a plunging wave and the resulting splash-up. For more examples of particle-based simulations of wave breaking and overtopping, read the works by Issa and Violeau,[37] Ma and Zhou[55] and Xie *et al.*[76]

6.6.2. *Multiphase flow*

Monaghan and Kocharyan[63] pioneered the first application of a particle method to multi-phase flows. In the field of MPS methods, Gotoh and Fredsøe[21] applied an MPS model for multi-phase simulations. Gotoh *et al.*[27] demonstrated the applicability of a multi-phase MPS method to simulate the detailed processes of wave generation due to landslides. Ikari *et al.*[36] applied a gas-liquid, two-phase MPS-based model to a plunging breaker. Gotoh *et al.*[23] and Gotoh and Sakai[26] provide detailed views on multi-phase MPS-based simulations.

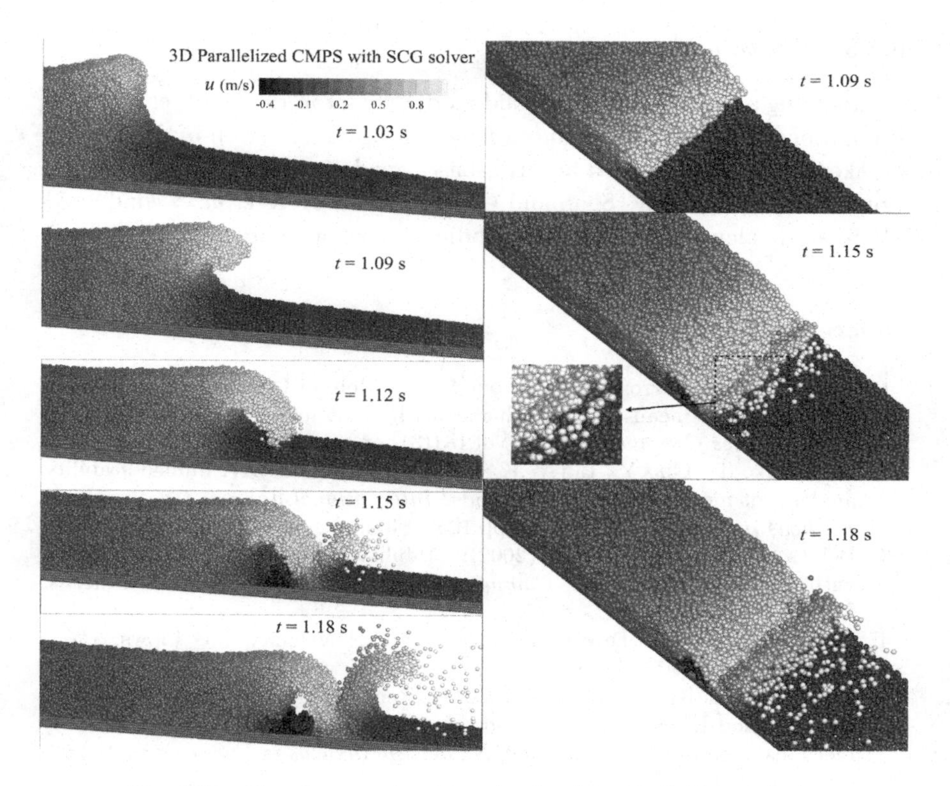

Fig. 6.12. Plunging wave breaking simulated by 3D CMPS method.

A limited number of SPH multiphase simulations are specifically oriented towards Coastal and Ocean Engineering. Violeau *et al.*[75] applied a multi-fluid SPH model to simulate the motion of a floating boom and an oil spill in a flume. Capone *et al.*[7] investigated wave generation processes induced by underwater landslides, where the landslide was modeled as a non-Newtonian fluid. Up to now, very few studies on particle-based simulations of multi-phase flows are encountered in Coastal and Ocean Engineering (e.g., aerated wave impacts on coastal structures, violent sloshing flows with air entrapment and entrainment, etc.). The paucity of studies is particularly due to the numerical difficulties with the mathematical discontinuity at the gas-liquid phase interface.

6.6.3. *Floating bodies*

By applying a passively moving solid model,[50] Gotoh *et al.*[28,29] carried out simulations of drift-timber induced flood and ice floes driven by a plunging breaker. Sueyoshi[73] applied an MPS-based model to simulate extreme motions of a floating body. Shao and Gotoh[67] and Rogers *et al.*[66] conducted SPH-based calculations of floating bodies in coastal zones.

References

1. Antuono M., Colagrossi A., Marrone S. and Molteni D. (2010): Free-surface flows solved by means of SPH schemes with numerical diffusive terms, *Computer Physics Communications*, Vol.181(3), pp. 532-549.
2. Belytschko T., Guo Y., Liu W.K. and Xiao S.P. (2000): A unified stability analysis of meshless particle methods, *International Journal for Numerical Methods in Engineering*, Vol.48, pp.1359-1400.
3. Belytschko T. and Xiao S.P. (2002): Stability analysis of particle methods with corrective derivatives, *Computers and Mathematics with Applications*, Vol.43, pp.329-350.
4. Bird G. A. (1987): Direct Simulation of High-Velocity Gas Flows, *Phys. Fluids*, Vol. 30, pp.364-366.
5. Bonet J. and Kulasegaram S. (2001): Remarks on tension instability of Eulerian and Lagrangian corrected smooth particle hydrodynamics (CSPH) methods, *International Journal of Numerical Methods in Engineering*, Vol.52, pp.1203-1220.
6. Bonet J. and Lok T. S. (1999): Variational and momentum preservation aspects of smooth particle hydrodynamic formulation, *Comput. Methods Appl. Mech. Eng.*, Vol.180, pp.97-115.
7. Capone T., Panizzo A. and Monaghan J.J. (2010): SPH modelling of water waves generated by submarine landslides, *J. Hydr. Res.*, Vol.48, pp.80-84.
8. Chen J.K., Beraun J.E. and Jih C.J. (1999): An improvement for tensile instability in smoothed particle hydrodynamics, *Comput. Mech.*, Vol.23, pp.279-287.
9. Colagrossi A.(2003): *A meshless Lagrangian method for free-surface and interface flows with fragmentation*, PhD Thesis, Universita di Roma, La Sapienza.
10. Colagrossi A. and Landrini M. (2003): Numerical simulation of interfacial flows by smoothed particle hydrodynamics, *Journal of Computational Physics*, Vol.191(2), pp.448-475.
11. Cooker M.J. and Peregrine D.H.(1992): Wave impact pressures and its effect upon bodies lying on the sea bed. *Coastal Engineering*, Vol.18 (3-4), pp.205-229.
12. Cundall P.A. and Strack O.D.L. (1979): A Discrete Numerical Model for Granular Assemblies, *Geotechnique*, Vol.29, pp.47-65.

13. Dalrymple R.A. and Rogers B.D. (2006): Numerical modeling of water waves with the SPH method, *Coastal Engineering*, Vol.53, pp.141-147.
14. Dilts G.A. (1999): Moving least squares hydrodynamics: Consistency and stability, *International Journal of Numerical Methods in Engineering*, Vol.44, pp.1115-1155.
15. Ellero M., Kroger M. and Hess S. (2002): Viscoelastic flows studied by smoothed particle dynamics, *Journal of Non-Newtonian Fluid Mechanics*, Vol.105(1), pp.35-51.
16. Fang J., Parriaux A., Rentschler M. and Ancey C. (2009): Improved SPH methods for simulating free surface flows of viscous fluids, *Applied Numerical Mathematics*, Vol.59(2), pp.251-271.
17. Fries T.P. and Belytschko T. (2008): Convergence and stabilization of stress-point integration in mesh-free and particle methods, *International Journal for Numerical Methods in Engineering*, Vol.74, pp.1067-1087.
18. Germano M., Piomelli U., Moin P. and Cabot W.H. (1991): A dynamic subgrid-scale eddy viscosity model, *Phys. Fluids*, A3, pp.1760-1765.
19. Gotoh H. (2004): *Computational Mechanics of sediment transport*, Morikita Shuppan Co., Ltd., p.223 (in Japanese).
20. Gotoh H. (2009): Lagrangian Particle Method as Advanced Technology for Numerical Wave Flume, *International Journal of Offshore and Polar Engineering*, Vol.19(3), pp.161-167.
21. Gotoh H. and Fredsøe J. (2000): Lagrangian two-phase flow model of the settling behavior of fine sediment dumped into water, *Proceedings of ICCE*, Sydney, Australia, pp.3906-3919.
22. Gotoh H., Hayashi M., Sakai T. and Oda K. (2003): Numerical model of wave breaking by Lagrangian particle method with Sub-Particle-Scale turbulence model, *Proc. APAC*, Makuhari, Japan.
23. Gotoh H., Ikari H., Memita T. and Sakai T. (2005): Lagrangian particle method for simulation of wave overtopping on a vertical seawall, *Coast. Eng. J.*, Vol.47, pp.157-181.
24. Gotoh H., Khayyer A., Ikari H. and Hori C. (2009): Refined reproduction of a plunging breaking wave and resultant splash-up by 3D-CMPS method, *Proc. 19th International Offshore and Polar Engineering, ISOPE*, Osaka, Japan, pp.518-524.
25. Gotoh H. and Sakai T. (1999): Lagrangian simulation of breaking waves using particle method, *Coast. Eng. J.*, Vol.41(3-4), pp.303-326.
26. Gotoh H. and Sakai T. (2006): Key issues in the particle method for computation of wave breaking, *Coastal Engineering*, Vol.53, pp.171-179.
27. Gotoh H., Sakai T. and Hayashi M. (2001): Lagrangian two-phase flow model for the wave generation process due to large-scale landslides, *Proc. APCE-2001*, Dalian, China, pp.176-185.
28. Gotoh H., Sakai T. and Hayashi M. (2002): Lagrangian model of drift-timbers induced flood by using moving particle semi-implicit method, *Journal of Hydroscience and Hydraulic Engineering* (JSCE), Vol.20(1), pp.95-102.
29. Gotoh H., Sakai T., Hayashi M. and Andoh S.(2002): Lagrangian gridless model for structure-flow-floats triangular interaction, *Proc. 13th IAHR-APD Cong.*, Singapore, Vol. 2, pp.765-770.

30. Gotoh H., Shibahara T. and Sakai T. (2001): Sub-Particle-Scale Turbulence Model for the MPS Method -Lagrangian Flow Model for Hydraulic Engineering, *Comput. Fluid Dynamics Jour.*, Vol.9(4), pp.339-347.
31. Gotoh H., Ikari H., Sakai T. and Mochizuki T. (2006) Development of numerical fishway by 3D MPS method, *Annual J. of Hyd. Eng.*, Vol. 50, pp.853-858 (in Japanese).
32. Gray J.P., Monaghan J.J. and Swift R.P. (2001): SPH elastic dynamics, *Comput. Methods Appl. Mech. Engrg.*, Vol.190, pp.6641-6662.
33. Hattori M., Arami A. and Yui T. (1994): Wave impact pressure on vertical walls under breaking waves of various types, *Coastal Engineering*, Vol.22 (1-2), pp.79-114.
34. Hori C., Gotoh H., Khayyer A. and Ikari H. (2011): Simulation of flip-through wave impact by CMPS method with SPS turbulence model, *Proc. Coastal Structures 2011*, Paper No. A8-26, Yokohama, Japan.
35. Hu X.Y. and Adams N.A. (2009): A constant-density approach for incompressible multi-phase SPH, *Journal of Computational Physics*, Vol.228(6), pp.2082-2091.
36. Ikari H., Gotoh H. and Sakai T. (2004): Simulation of wave breaking by the particle method with liquid-gas two-phase-flow model, *Annual Journal of Coastal Engineering* (JSCE), pp.111-115 (in Japanese).
37. Issa R. and Violeau D. (2008): Modelling a plunging breaking solitary wave with eddy-viscosity turbulent SPH models. *Computers, Materials, & Continua*, Vol.8(3), pp.151-164.
38. Khayyer A. and Gotoh H. (2008): Development of CMPS method for accurate water-surface tracking in breaking waves, *Coast. Eng. J.*, Vol.50(2), pp.179-207.
39. Khayyer A. and Gotoh H. (2009): Modified Moving Particle Semi-implicit methods for the prediction of 2D wave impact pressure, *Coastal Engineering*, Vol.56, pp.419-440.
40. Khayyer A. and Gotoh H. (2010): A Higher Order Laplacian Model for Enhancement and Stabilization of Pressure Calculation by the MPS Method, *Applied Ocean Res.*, Vol.32(1), pp.124-131.
41. Khayyer A., Gotoh H. and Shao S.D. (2008): Corrected Incompressible SPH method for accurate water-surface tracking in breaking waves, *Coastal Engineering*, Vol.55(3), pp.236-250.
42. Khayyer A., Gotoh H. and Shao S.D. (2009): Enhanced predictions of wave impact pressure by improved incompressible SPH methods, *Applied Ocean Research*, Vol.31(2), pp.111-131.
43. Khayyer A. and Gotoh H. (2011): Enhancement of Stability and Accuracy of the Moving Particle Semi-implicit Method, *Journal of Computational Physics*, Vol.230, pp.3093-3118.
44. Kishev Z.R., Hu C. and Kashiwagi M. (2006): Numerical simulation of violent sloshing by a CIP-based method, *Journal of Marine Science and Technology*, Vol.11(2), pp.111-122.
45. Kondo M. and Koshizuka S. (2011): Improvement of Stability in Moving Particle Semi-implicit method, *International Journal for Numerical Methods in Fluids*, Vol.65, pp.638-654.

46. Kondo M., Suzuki Y. and Koshizuka S. (2010): Suppressing local particle oscillations in the Hamiltonian particle method for elasticity, *International Journal for Numerical Methods in Engineering*, Vol.81(12), pp.1514-1528.
47. Koshizuka S. (2008): *Particle Method Simulation*, Baifukan Co., Ltd., p.179 (in Japanese).
48. Koshizuka S. and Oka Y. (1996): Moving particle semi-implicit method for fragmentation of incompressible fluid, *Nuclear Science and Engineering*, Vol.123, pp.421-434.
49. Koshizuka S., Ikeda H. and Oka Y. (1999): Numerical analysis of fragmentation mechanisms in vapor explosions, *Nuclear Engineering and Design*, Vol.189, pp.423-433.
50. Koshizuka S., Nobe A. and Oka Y. (1998): Numerical Analysis of Breaking Waves Using the Moving Particle Semi-implicit Method, *Int. J. Numer. Meth. Fluid*, Vol.26, pp.751-769.
51. Koshizuka S. (2005): *Particle Method*, Maruzen Publishing, p.144 (in Japanese).
52. Lee E.S., Moulinec C., Xu R., Violeau D., Laurence D. and Stansby P. (2008): Comparisons of weakly-compressible and truly incompressible algorithms for the SPH mesh free particle method, *Journal of Computational Physics*, Vol.227(18), pp.8417-8436.
53. Li Y. and Raichlen F. (2003): Energy balance model for breaking solitary wave runup, *Journal of Waterway, Port, Coastal, and Ocean Engineering*, Vol.129(2), pp.47-59.
54. Lucy L. B. (1977): A numerical approach to the testing of the fission hypothesis, *Astron. J.*, Vol.82, pp.1013-1024.
55. Ma Q.W. and Zhou J.T. (2009): MLPG-R Method for Numerical Simulation of 2D Breaking Waves, *Computer Modeling in Engineering & Sciences* (CMES), Vol.43(3), pp.277-304.
56. McNamara G. R. and Zanetti G. (1988): Use of Boltzmann Equation to Simulate Lattice-Gas Automata, *Phys. Rev. Lett.*,Vol. 61, pp.2332-2335.
57. Melean Y., Di L., Sigalotti G. and Hasmy A. (2004): On the SPH tensile instability in forming viscous liquid drops, *Computer Physics Communications*, Vol.157, p.191.
58. Molteni D. and Colagrossi A. (2009): A simple procedure to improve the pressure evaluation in hydrodynamic context using the SPH, *Computer Physics Communications*, Vol.180(6), pp.861-872.
59. Monaghan J.J. (1992): Smoothed particle hydrodynamics, *Ann. Rev. Astron. Astrophys*, Vol.30, pp.543-574.
60. Monaghan J.J. (1994): Simulating free surface flows with SPH, *Journal of Computational Physics*, Vol.110, pp.399-406.
61. Monaghan J.J. (2000): SPH without a tensile instability, *Journal of Computational Physics*, Vol.159(2), pp.290-311.
62. Monaghan J.J. (2005): Smoothed particle hydrodynamics, *Reports on Progress in Physics*, Vol.68, pp.1703-1759.
63. Monaghan J.J. and Kocharyan A. (1995): SPH simulation of multi-phase flow, *Computer Physics Communications*, Vol.87, pp.225-235.

64. Morris J.P., Fox P.J. and Zhu Y. (1997): Modeling low Reynolds number incompressible flows using SPH, *Journal of Computational Physics*, Vol.136, pp.214-226.
65. Oger G., Doring M., Alessandrini B. and Ferrant P. (2007): An improved SPH method: Towards higher order convergence, *Journal of Computational Physics*, Vol.225(2), pp.1472-1492.
66. Rogers B., Dalrymple R.A., Stansby P. (2008): SPH Modeling of floating bodies in the surf zone, *Proc. 31st In. Conf. Coastal Eng*, Hamburg, Germany, pp.204-215.
67. Shao S.D. and Gotoh H. (2004): Simulation coupled motion of progressive wave and floating curtain wall by SPH-LES model, *Coast. Eng. J.*, Vol.46(2), pp.171-202.
68. Shao S.D., Ji C.M., Graham D.I., Reeve D.E., James P.W. and Chadwick A.J. (2006): Simulation of wave overtopping by an incompressible SPH model, *Coastal Engineering*, Vol.53(9), pp.723-735.
69. Shao S.D. and Lo E.Y.M. (2003): Incompressible SPH method for simulating Newtonian and non-Newtonian flows with a free surface, *Advanced Water Resources*, Vol.26 (7), pp.787-800.
70. Shibata K. and Koshizuka S. (2007): Numerical analysis of shipping water impact on a deck using a particle method, *Ocean Engineering*, Vol.34, pp.585-593.
71. Sigalotti L.D.G. and Lopez H. (2008): Adaptive kernel estimation and SPH tensile instability, *Computers and Mathematics with Applications*, Vol.55(1), pp.23-50.
72. Smagorinsky J. (1963): General circulation experiments with the primitive equations, I. The basic experiment, *Monthly Weather Review*, Vol.91, pp.99-164.
73. Sueyoshi M. (2004): Numerical simulation of extreme motions of a floating body by MPS method, *Proc. of Joint Int. Conf. of OCEANS'04 and TECHNO-OCEAN'04*, pp. 566-572.
74. Swegle J.W., Hicks D.L. and Attaway S.W. (1995): Smoothed Particle Hydrodynamics Stability Analysis, *Journal of Computational Physics*, Vol.116, p.123.
75. Violeau D., Buvat C., Abed-Meraim K. and de Nanteuil E. (2007): Numerical modelling of boom and oil spill with SPH Original Research Article, *Coastal Engineering*, Vol.54(12), pp.895-913.
76. Xie J., Nistor I. and Murty T. (2012): A corrected 3-D SPH method for breaking tsunami wave modelling, *Natural Hazards*, Vol.60(1), pp.81-100.

Chapter 7

Distinct Element Method

Eiji Harada and Hitoshi Gotoh

Cundall and Strack[2] proposed the Distinct Element Method (DEM), which is categorized as a particle method in a broad sense. Since the 1990s, the DEM has been applied to coastal engineering subjects such as sediment transport,[1,3,5,10] sedimentation processes in reclamation work[11,18] and stability of block mounds,[9,14,15] and the mechanisms of these phenomena.

In this chapter, the computing method of the DEM, which contains the governing equations, coordinate transformation and a tuning of model parameters, is described in detail. As a tracking tool of armor blocks, extended DEM to simulate rigid body motion with a quaternion description is also explained.

7.1. Computing Method

7.1.1. *Governing equations*

In the DEM, a spring-dashpot system is introduced to evaluate the contact force between elements, and the behavior of the assembled elements is computed by numerically integrating the translational and rotational equations of motion for each element. For simplicity, a circle (two-dimensional computation) or a sphere (three-dimensional computation) is used to develop a simple algorithm to detect neighboring elements.

Each individual element is governed by the translational and the rotational equations of motion. The hydrodynamic force acting on individual elements and the inter-element forces due to collisions between elements must be taken into account as follows:

$$M\frac{d\boldsymbol{v}}{dt} = \boldsymbol{F}_{pint} + \boldsymbol{F}_{flow} + \boldsymbol{F}_g \tag{7.1}$$

$$\boldsymbol{I}\frac{d\boldsymbol{\omega}}{dt} = \boldsymbol{T}_{pint} + \boldsymbol{T}_{flow} \tag{7.2}$$

where M is the mass of each individual element, $\boldsymbol{v}$ is the velocity of each element, $\boldsymbol{F}_{pint}$ is the inter-element force, $\boldsymbol{F}_{flow}$ is the hydrodynamic force, $\boldsymbol{F}_g$ is the body force, $\boldsymbol{I}$ is the inertia tensor, $\boldsymbol{\omega}$ is the angular velocity of each element, $\boldsymbol{T}_{pint}$ is the torque due to the inter-element force, and $\boldsymbol{T}_{flow}$ is the torque due to the hydrodynamic force.

The hydrodynamic force can be computed by integrating the fluid stress tensor over the surface of each element as follows:

$$\boldsymbol{F}_{flow} = \iint_s \boldsymbol{\tau} \cdot \boldsymbol{n}\, ds \tag{7.3}$$

$$\boldsymbol{T}_{flow} = \iint_s \boldsymbol{r}_s \times (\boldsymbol{\tau} \cdot \boldsymbol{n})\, ds \tag{7.4}$$

in which $\boldsymbol{n}$ is the unit normal vector directed outward from the surface of the element, $\boldsymbol{\tau}$ is the fluid stress tensor, s is the surface of the element, and $\boldsymbol{r}_s$ is the relative positional vector from the centroid of the element to its surface. Note that this kind of the computation requires a fine computational grid to resolve eddies on the Kolmogorov scale, which consequently yields an extremely high computational load. Therefore, an empirical formula is usually employed to describe the hydrodynamic force. For example, the empirical formula for the drag force is described as follows:

$$\boldsymbol{F}_{flow} = \frac{1}{2}\rho C_D A_2 d^2 \,|\,\boldsymbol{u} - \boldsymbol{v}\,|\,(\boldsymbol{u} - \boldsymbol{v}) \tag{7.5}$$

where C_D is the drag coefficient, ρ is the fluid density, A_2 is the two dimensional geometrical coefficient of the element, d is the diameter of the element and $\boldsymbol{u}$ is the fluid velocity. Usually, the torque due to the hydrodynamic force is neglected for simplicity.

If the relative distance between the centroids of elements i and j satisfies

$$|\,\boldsymbol{r}_{i,j}\,| \le \frac{d_i + d_j}{2} \tag{7.6}$$

the inter-element force is activated, where $\boldsymbol{r}_{i,j}$ is the relative positional vector between elements i and j, and d_i is the diameter of element i. The inter-element force is calculated with mechanical joints (see Fig. 7.1), which are arranged in both normal and tangential directions on the tangential plane having the common contact point on elements i and j. The relative displacement that overlaps between elements i and j on the global coordinate O-xyz can be written as

$$\Delta \boldsymbol{x}_{i,j} = \left[(\boldsymbol{v} + \boldsymbol{\omega} \times \boldsymbol{r})\, \Delta t\right]_{i,j} \tag{7.7}$$

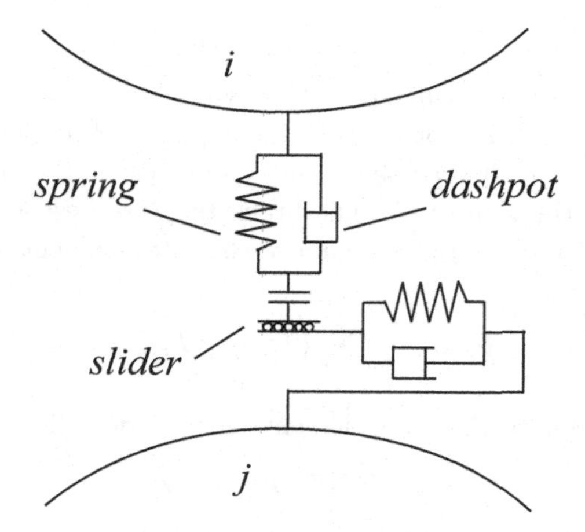

Fig. 7.1. Spring–dashpot system.

where $\boldsymbol{r}$ is the relative positional vector from the centroid of the element to its surface, Δt is the time marching step. And on the local coordinate O-$\xi\eta\zeta$ the relative displacement can be written as

$$\Delta\boldsymbol{\xi}_{i,j} = \boldsymbol{T}_{GL} \cdot \Delta\boldsymbol{x}_{i,j} \tag{7.8}$$

where $\boldsymbol{T}_{GL}$ is the transformation matrix from the global coordinate O-xyz to the local coordinate O-$\xi\eta\zeta$. A square bracket $[\ ]_{i,j}$ in Eq. (7.7) indicates the sign of the relative displacement that overlaps elements i and j. The relative displacement due to rotational motion is neglected in Eq.(7.8).

The mechanical joint is composed of a spring, dashpot and friction slider as shown in Fig. 7.1. The inter-element force is written as follows:

$$\boldsymbol{F}_{pint,\,L} = \boldsymbol{e} + \boldsymbol{d}; \quad \boldsymbol{e} = \boldsymbol{e}^{pre} + k\Delta\boldsymbol{\xi}_{i,j}\,; \quad \boldsymbol{d} = c\Delta\dot{\boldsymbol{\xi}}_{i,j} \tag{7.9}$$

where $\boldsymbol{e}$ is the inter-element force due to the spring, $\boldsymbol{d}$ is the inter-element force due to the dashpot, k is the stiffness, and c is the damping coefficient. The superscript "*pre*" denotes the previous time step, and the subscript "L" denotes the local coordinate O-$\xi\eta\zeta$. Moreover, a joint without resistance to tensile forces is introduced, and a slider joint is placed to represent frictional effects. Thus according to the contact condition between elements, the inter-element force $\boldsymbol{F}_{pint,L}$ given by Eq. (7.9) is modified as

$$\langle e_{\perp} \rangle < 0 \quad \text{then} \quad \boldsymbol{F}_{pint,\,L} = \boldsymbol{0} \tag{7.10}$$

$$\left|\langle e_{||}\rangle\right| > \mu \langle e_{\perp}\rangle \quad \text{then} \quad \langle \boldsymbol{F}_{pint,\,L,\,||}\rangle = \mu \cdot \mathrm{Sign}\left[\langle e_{\perp}\rangle, \langle e_{||}\rangle\right] \tag{7.11}$$

where μ is the friction coefficient. The symbols "$\perp$" and "$||$" denote the vector components in the normal and the tangential directions, respectively. In addition, Sign$[A, B]$ represents the operation that gives the absolute value of A with the sign of B. The brackets $\langle\ \rangle$ represent the component of the vector. Consequently, the sum of the inter-element forces acting on element i is given as

$$\boldsymbol{F}_{pint} = -\sum_{j} \left(\boldsymbol{T}_{GL}\right)^{-1} \boldsymbol{F}_{pint,\,L} \tag{7.12}$$

and the torque due to the inter-element forces is described by

$$\boldsymbol{T}_{pint} = -\frac{d}{2}\sum_{j} \left(\boldsymbol{T'}_{GL}\right)^{-1} \boldsymbol{F}_{pint,\,L} \tag{7.13}$$

In these descriptions, the torsional moment about the normal direction for three-dimensional simulations is omitted. The time evolution of the motion of each element can be computed using an appropriate time differential scheme such as the Euler explicit scheme.

7.1.2. *Coordinate transformation*

The η-axis in the local coordinate O-$\xi\eta\zeta$ is defined to be on the xy-plane. The transformation from the global coordinate O-xyz to the local coordinate O-$\xi\eta\zeta$ is shown in Fig. 7.2 and can be expressed by the transformation matrix

$$\begin{aligned} \boldsymbol{T}_{GL} &= \boldsymbol{R}_{\eta}\boldsymbol{R}_{z} \\ &= \begin{bmatrix} \cos\gamma & 0 & -\sin\gamma \\ 0 & 1 & 0 \\ \sin\gamma & 0 & \cos\gamma \end{bmatrix} \begin{bmatrix} \cos\beta & \sin\beta & 0 \\ -\sin\beta & \cos\beta & 0 \\ 0 & 0 & 1 \end{bmatrix} \\ &= \begin{bmatrix} \cos\gamma\cos\beta & \cos\gamma\sin\beta & -\sin\gamma \\ -\sin\beta & \cos\beta & 0 \\ \sin\gamma\cos\beta & \sin\gamma\sin\beta & \cos\gamma \end{bmatrix} \end{aligned} \tag{7.14}$$

in which $\boldsymbol{R}_*$ is the rotation matrix around the *-axis, and β, γ are rotational angles around the z-axis and η-axis, respectively. This transformation matrix can be rewritten with the direction cosines (l, m, n)

$$\begin{aligned} l &= \frac{x_j - x_i}{L_{ij}};\ m = \frac{y_j - y_i}{L_{ij}};\ n = \frac{z_j - z_i}{L_{ij}} \\ L_{ij} &= \sqrt{(x_j - x_i)^2 + (y_j - y_i)^2 + (z_j - z_i)^2} \end{aligned} \tag{7.15}$$

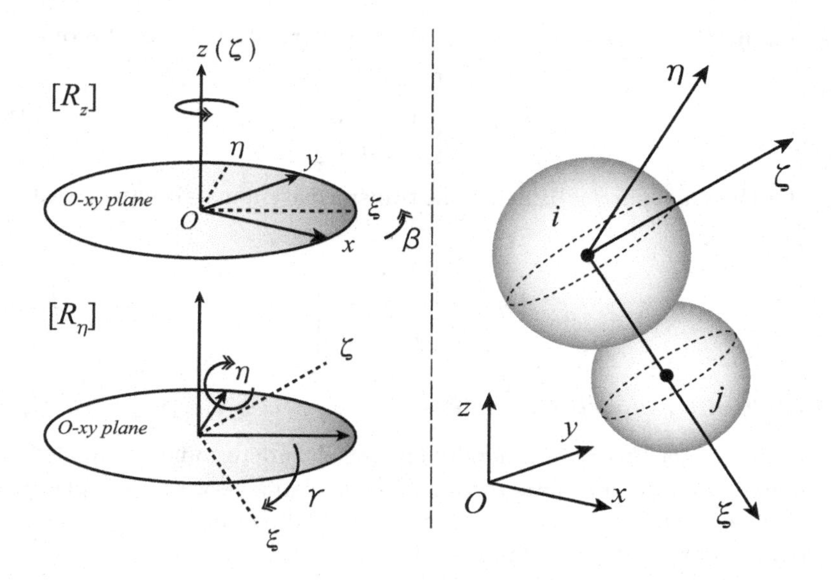

Fig. 7.2. Global and local coordinate systems.

which gives the relation between the ξ-axis and the global coordinate O-xyz, as follows:

$$\boldsymbol{T}_{GL} = \begin{bmatrix} l & m & n \\ -\dfrac{m}{\pm\sqrt{l^2+m^2}} & \dfrac{l}{\pm\sqrt{l^2+m^2}} & 0 \\ -\dfrac{ln}{\pm\sqrt{l^2+m^2}} & -\dfrac{mn}{\pm\sqrt{l^2+m^2}} & \pm\sqrt{l^2+m^2} \end{bmatrix} \tag{7.16}$$

This transformation matrix has two alternatives because $\cos\gamma$ can have a positive or negative sign. For the case when $l = m = 0$ (i.e., the x- and y-coordinates of element i correspond to those of element j), the denominator in the matrix $\sqrt{l^2+m^2}$ is zero. For this case, considering that $\cos\gamma=0$, Eq. (7.14) is expressed as follows:

$$\mathbf{T}_{GL} = \begin{bmatrix} 0 & 0 & n \\ -\sin\beta & \cos\beta & 0 \\ -n\cos\beta & -n\sin\beta & 0 \end{bmatrix} \tag{7.17}$$

Since the angle β between the η-axis and the y-axis does not affect the relative displacement of the z- or ξ-axis, the angle β can be chosen arbitrarily.

For simplicity, for the case of $\beta = 0$, the transformation matrix becomes

$$\mathbf{T}_{GL} = \begin{bmatrix} 0 & 0 & n \\ 0 & 1 & 0 \\ -n & 0 & 0 \end{bmatrix} \tag{7.18}$$

Also for this case, $\sin\gamma = 1$, hence the transformation matrix can be written as

$$\mathbf{T}_{GL} = \begin{bmatrix} 0 & 0 & \pm 1 \\ 0 & 1 & 0 \\ 1 & 0 & 0 \end{bmatrix} \tag{7.19}$$

7.1.3. *Tuning model parameters*

Hertzian contact theory is generally employed to determine the model parameters.[6] In Hertzian contact theory, the overlap δ between two spheres is determined by the contact force P in the normal direction, Young's modulus E, and Poisson's ratio ν as follows:

$$\delta^3 = \frac{9}{2}\frac{d_i + d_j}{d_i d_j}\left(\frac{1-\nu^2}{E}\right)^2 P^2 \tag{7.20}$$

in which d_i is the diameter of particle i (see Fig. 7.3). Consequently, the non-linear relation between the contact force and overlap is given as follows:

$$P = k_\perp \delta^{3/2} \tag{7.21}$$

$$k_\perp = \frac{\sqrt{2}}{3}\sqrt{\frac{d_1 d_2}{d_1 + d_2}}\frac{E}{1-\nu^2} \tag{7.22}$$

in which $k_\perp$ is the spring constant in the normal direction. On the other hand, the relation between contact force and overlap can be described as a pseudo linear relation:

$$P = k_\perp \delta \tag{7.23}$$

$$k_\perp = \left\{\frac{2}{9}\frac{d_1 d_2}{d_1 + d_2}\left(\frac{E}{1-\nu^2}\right)^2 \cdot P\right\}^{1/3} \tag{7.24}$$

Here the spring constant in the normal direction can be estimated with the normal force (see Eqs. (7.9) and (7.10)):

$$P = \langle e_\perp \rangle \tag{7.25}$$

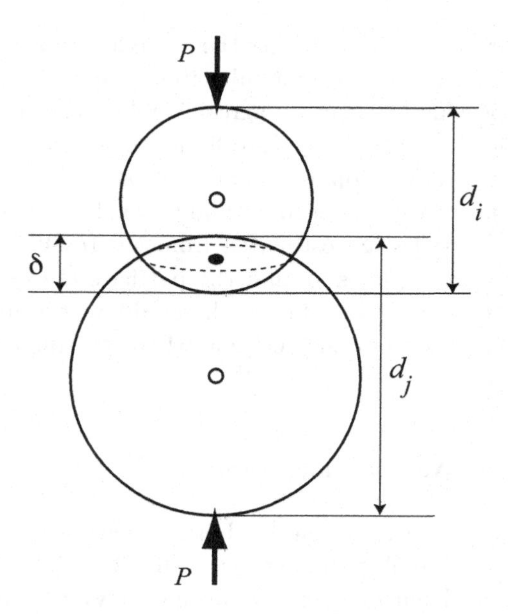

Fig. 7.3. Contact between two particles.

The spring constant $k_{||}$ is estimated using the ratio of Young's modulus E to the shear modulus G as follows:

$$s_0 = \frac{k_{||}}{k_\perp} = \frac{G}{E} = \frac{1}{2\,(1+\nu)} \tag{7.26}$$

where s_0 is the damping rate. Then, viscosity constants $c_\perp$ and $c_{||}$ are deduced from the critical damping condition of the Voigt model with a single degree of freedom.

$$c_\perp = 2\sqrt{Mk_\perp}\ ;\ c_{||} = c_\perp\sqrt{s_0} \tag{7.27}$$

where M is the mass of the particle. Considering numerical stability, the time increment of computation Δt should be set sufficiently small. The time increment of computation can be selected using the relation of the natural period T in the spring-mass system as follows:

$$\Delta t = \frac{T}{\alpha} \tag{7.28}$$

$$T = 2\pi\sqrt{\frac{M}{k_\perp}} \tag{7.29}$$

in which α is a parameter where $10 \leq \alpha \leq 20$.

The tuning process above indicates the key role of the Young's modulus E. Note that estimating Young's modulus from a precise material test does not necessarily provide satisfactory agreement between numerical simulations and experiments. Since there are limits to computational resources, it is impossible to track a copious number of particles while maintaining sufficient resolution of the flow field around them. In most cases, hydraulic forces acting on particles are estimated from low-resolution velocity flow fields. In other words, eddies around the particles are not well resolved. To account for the energy-loss due to these eddies, the mechanism must be empirically modeled, and one such model treats Young's modulus as a tuning parameter.

7.2. DEM-based Armor Block Model

To simulate how an individual armor block behaves under the onslaught of large waves, the complex shape of an individual armor unit has to be represented in a rational manner. A convenient way to approximate the geometry of an armor block entails the linking of some standard elements, such as circles and spheres, for two- and three-dimensional simulations, respectively. In this section, first we explain the DEM-based armor block model. Then we present three-dimensional DEM simulation results of the compaction process that occurs within the wave-dissipating armor block layer of an offshore breakwater.

7.2.1. *Rigid body motion*

The behavior of an individual armor block is usually tracked using a rigid body model. The passively moving solid model proposed by Koshizuka *et al.*[13] has been frequently employed as a rigid body model. This model was developed as a rigid body model using the Moving Particle Semi-implicit (MPS) method.[12] Because both DEM and MPS method have common characteristics, such as being Lagrangian models, the passively moving solid model shows high affinity with the DEM.

In the passively moving solid model, the motions of all elements that compose an individual armor block are calculated first using the DEM without rigid connections. Consequently, as shown in Fig. 7.4, the relative positions of the individual armor block elements are allowed to vary.

With these auxiliary positional and velocity vector components, the translational velocity vector $\boldsymbol{v}_g$ and rotational velocity vector $\boldsymbol{\omega}_g$ acting on

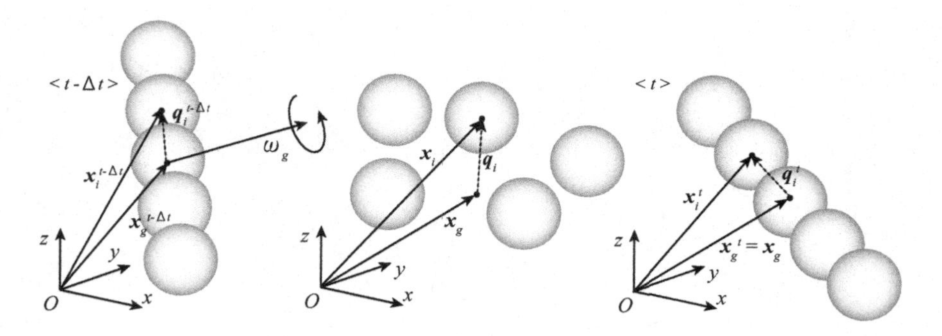

Fig. 7.4. Coordinate modification for armor block components.

each individual armor block centroid $\boldsymbol{x}_g$ are given as follows:

$$\boldsymbol{v}_g = \frac{1}{k}\sum_{i=1}^{k} \boldsymbol{v}_i \tag{7.30}$$

$$\boldsymbol{\omega}_g = \boldsymbol{I}_g^{-1}\sum_{i=1}^{k} (\boldsymbol{q}_i \times \boldsymbol{v}_i) \tag{7.31}$$

$$\boldsymbol{x}_g = \frac{1}{k}\sum_{i=1}^{k} \boldsymbol{x}_i \; ; \; \boldsymbol{q}_i = \boldsymbol{x}_i - \boldsymbol{x}_g \tag{7.32}$$

in which $\boldsymbol{x}_i$ is the auxiliary positional vector of the i-th component of the armor block, $\boldsymbol{v}_i$ is the auxiliary velocity vector of the i-th component of the armor block, k is the number of components of the armor block, and $\boldsymbol{I}_g$ is the inertia tensor of the armor block. Then the positional vector of the armor block after a specified time is given by

$$\boldsymbol{x}_i^t = \boldsymbol{x}_i^{t-\Delta t} + \left(\boldsymbol{v}_g + \boldsymbol{\omega}_g \times \boldsymbol{q}_i^{t-\Delta t}\right)\Delta t \tag{7.33}$$

By the way, it should be noted that using the correction in Eq. (7.33) yields gaps between block components due to the numerical error shown in Fig. 7.5 (a); gap generation is especially remarkable in the uniform rotation under a large time marching step. To appropriately address this numerical error, we introduce a quaternion description based on the rotational angle (see Fig. 7.5 (b)). The quaternion is commonly used in the field of computer graphics. By quaternion representation, the relative positional vector to the

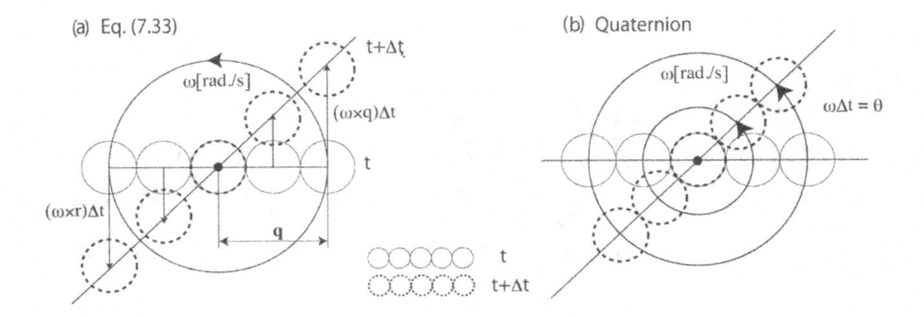

Fig. 7.5. Correction for rotational motion.

armor block centroid at time t is given by

$$Q_i^t = \left(0, \boldsymbol{q}_i^t\right) \tag{7.34}$$

Additionally, this quaternion has the following relation with its quaternion at time t-Δt:

$$Q_i^t = QQ_i^{t-\Delta t}Q^* = Q\left(0, \boldsymbol{q}_i^{t-\Delta t}\right)Q^* \tag{7.35}$$

in which Q^* is the conjugate quaternion. A quaternion consists of four components as follows:

$$Q = (q_0, \boldsymbol{q}) = (q_0, q_x, q_y, q_z) \tag{7.36}$$

And its conjugate quaternion is defined as

$$Q^* = (q_0, -\boldsymbol{q}) \tag{7.37}$$

A quaternion that represents the rotation around the axis $\boldsymbol{\Psi} = \boldsymbol{\omega}_g/|\boldsymbol{\omega}_g|$ is described as follows:

$$Q = \left(\cos\frac{\theta}{2}, \Psi_x \sin\frac{\theta}{2}, \Psi_y \sin\frac{\theta}{2}, \Psi_z \sin\frac{\theta}{2}\right) \quad ; \quad \theta = |\boldsymbol{\omega}_g|\,\Delta t \tag{7.38}$$

Using the components of quaternion, the rotation matrix can be written as

$$\boldsymbol{R} = \begin{bmatrix} 1-2q_y^2-2q_z^2 & 2q_xq_y-2q_0q_z & 2q_xq_z+2q_0q_y \\ 2q_xq_y+2q_0q_z & 1-2q_x^2-2q_z^2 & 2q_yq_z-2q_0q_x \\ 2q_xq_z-2q_0q_y & 2q_yq_z+2q_0q_x & 1-2q_x^2-2q_y^2 \end{bmatrix} \tag{7.39}$$

Consequently, the quaternion representing the relative positional vector to the armor block centroid at time t is written as

$$Q_i^t = \left(0, \boldsymbol{R}\boldsymbol{q}_i^{t-\Delta t}\right) \tag{7.40}$$

Thus, the positional vectors of the components of the armor block $\boldsymbol{x}_i^t$ are given by

$$\boldsymbol{x}_i^t = \boldsymbol{x}_g + \boldsymbol{q}_i^t = \boldsymbol{x}_g + \boldsymbol{R}\boldsymbol{q}_i^{t-\Delta t} \quad (7.41)$$

Gotoh and Ikari[8] simulated a girder bridge during a tsunami run-up, modeling the driftwood and tsunami run-up using the MPS method. They modeled the bridge and driftwood as rigid bodies composed of particles and applied the passively moving solid model.[13] In their simulation, they applied the quaternion to calculate the positional vector of the rigid body with rotational motion in the same manner as Shibata *et al.*[16]

7.2.2. *Tracking armor blocks under high waves*

In this section the compaction process of wave-dissipating blocks under the repeated action of high waves is shown as an example of the DEM-based block model mentioned above.

Offshore breakwaters are exposed to severe high waves during a storm, enduring strong wave forces repeatedly. Wave-dissipating blocks located on the front of the breakwater often show a remarkable subsidence. Some reasons for the subsidence include: (i) Liquefaction of the seabed under the wave-dissipating blocks;[19] (ii) Local scouring at the toe of the wave-dissipating block mound;[17] and (iii) Compaction of the wave-dissipating blocks with the change of engagement between blocks.[9]

Gotoh *et al.*[7] used a three-dimensional DEM-based armor block model to show that an irregular disposition of blocks greatly influenced the compaction process. Harada *et al.*[9] conducted hydraulic experiments of the block compaction process caused by the change of engagement between blocks under the action of high waves. They also reproduced the compaction process using the same 3D-DEM-based block model as Gotoh *et al.*[7] and found good agreement with the experimental results as shown in Fig. 7.6. Although the wave forces exerted on the blocks were simply evaluated with Goda's formula,[4] acceptable agreement between the experiment and simulation was found. Additionally, this DEM-based block model is a useful tool to investigate the internal structure and interaction forces between blocks during the compaction process.

The three-dimensional DEM-based block model, combined with a numerical wave flume to reproduce three-dimensional free surface flows, will provide an ideal tool for computer-aided design of coastal structures.

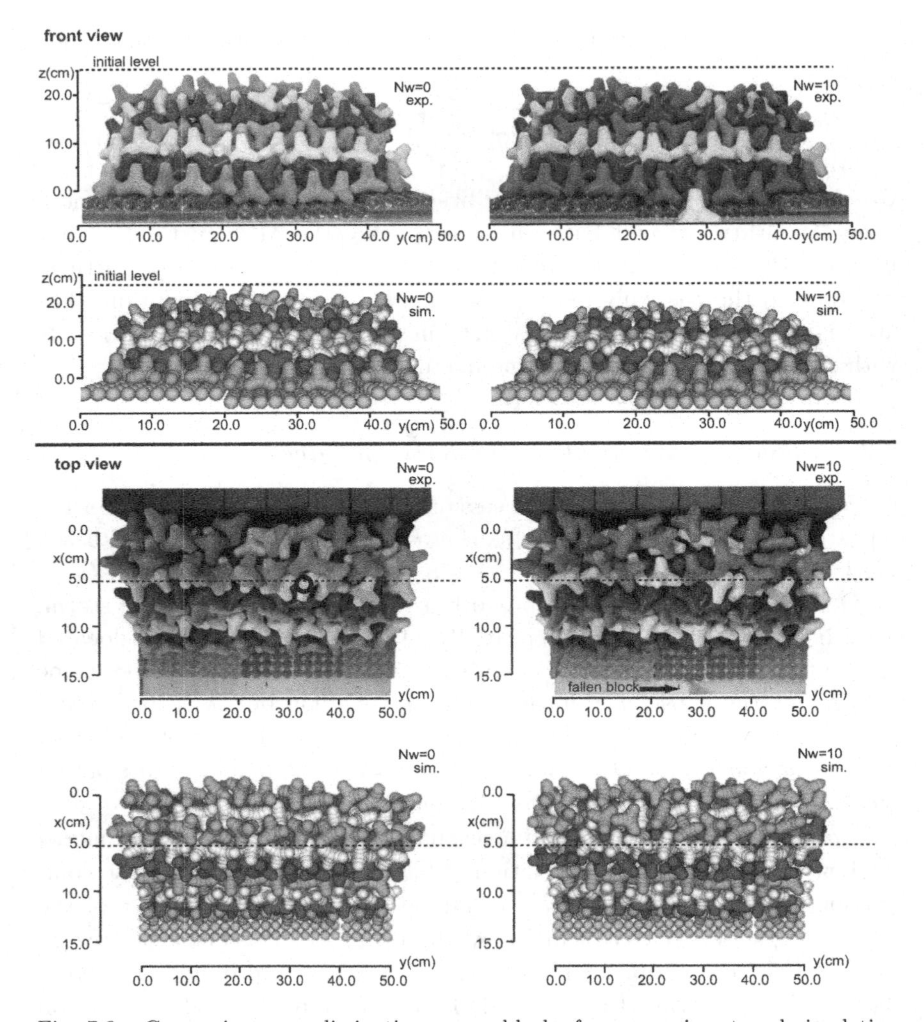

Fig. 7.6. Comparing wave-dissipating armor blocks from experiment and simulation (N_w: number of waves).

References

1. Calantoni J. and Thaxton C.S. (2008): Simple Power Law for Transport Ratio with Bimodal Distributions of Coarse Sediments under Wave, *J. Geophys. Res.*, Vol.113, C03003.
2. Cundall P.A. and Strack O.D.L. (1979): A Discrete Numerical Model for Granular Assemblies, *Geotechnique*, Vol.29, pp.47-65.

3. Drake T.G. and Calantoni J. (2001): Discrete particle model for sheet flow sediment transport in the nearshore, *J. Geophys. Res.*, Vol.106(C9), pp.19859-19868.
4. Goda Y. (1973): A new method of wave pressure calculation for the design of composite breakwaters, *Report of the Port and Harbour Research Institute*, Vol.12, No.3, pp.31-69 (in Japanese).
5. Gotoh H. and Sakai T. (1997): Numerical simulation of sheetflow as granular material, *Journal of Waterway, Port, Coastal, and Ocean Engineering*, Vol.123, No.6, pp.329-336.
6. Gotoh H. (2004): *Computational Mechanics of Sediment Transport*, Morikita Shuppan Co., Ltd. (in Japanese).
7. Gotoh H., Harada E., Takayama T., Mizutani M., Fudou M. and Iwamoto T. (2005): Compaction mechanism of wave dissipating blocks due to high waves, *Annual Jour. of Coastal Engineering*, JSCE, Vol.52, pp.781-785 (in Japanese).
8. Gotoh H. and Ikari H. (2007): Numerical analysis on girder bridge washed away by tsunami run-up, *Proceedings of International Conference on Violent Flows*, pp.159-164.
9. Harada E., Gotoh H. and Sakai T. (2007): 3D Lagrangian simulation of compaction process of wave dissipating blocks due to high waves, *Proceedings of International Conference on Violent Flows*, pp.221-226.
10. Harada E. and Gotoh H. (2008): Computational mechanics of vertical sorting of sediment in sheetflow regime by 3D granular material model, *Coast. Eng. J.*, Vol.50, No.1, pp.19-45.
11. Harada E., Gotoh H. and Tsuruta N. (2011): Numerical simulation for sedimentation process of blocks on a sea bed by high-resolution multi-phase model, *Coast. Eng. J.*, Vol.53, No.4, pp.343-364.
12. Koshizuka S. and Oka Y. (1996): Moving-particle semi-implicit method for fragmentation of incompressible fluid, *Nuclear Science and Engineering*, Vol.123, pp.421-434.
13. Koshizuka S., Nobe A. and Oka Y. (1998): Numerical analysis of breaking waves using the moving particle semi-implicit mothod, *Int. J. Numer. Mech. Fluids*, Vol.26, pp.751-769.
14. Latham J.-P., Munjiza A., Mindel J., Xiang J., Guises R., Garcia X., Pain C., Gorman G. and Piggott. M. (2008): Moedelling of massive particulates for breakwater engineering using coupled FEMDEM and CFD, *Particuology*, Vol.6, pp.572-583.
15. Latham J.-P., Mindel J., Xiang J., Guises R., Garcia X., Pain C., Gorman G., Piggott M. and Munjiza A. (2009): Coupled FEMDEM/Fluids for coastal engineers with special reference to armour stablility and breakage, *Geomechanics and Geoengineering*, Vol.4, No.1, pp.39-53.
16. Shibata K., Koshizuka S., Oka Y. and Tanizawa K. (2004): A three-dimensional numerical analysis code for shipping water on deck using a particle method, *Proceedings of ASME 2004 Heat Transfer/Fluids Engineering Summer Conference*, pp. 959-964.

17. Suzuki K., Takahashi S., Takano T. and Shimosako K. (2002): Settlement failure of wave dissipating blocks in front of caisson type breakwater due to scouring under the rubble mound -Field investigation and large scale experiment-, *Report of the Port and Airport Research Institute*, Vol.41, No.1, pp.51-53, 55-89 (in Japanese).
18. Oda K. and Shigematsu T. (1994): Development of a numerical simulation method for predicting the settling behavior and deposition configuration of soil dumped into waters, *Proceeding of ICCE*, Kobe, Japan, pp.3305-3319.
19. Zen K. and Yamazaki H. (1990): Deformation of rubble sloping breakwater due to the wave-induced liquefaction in foundation, *Proc. of Civil Engineering in the Ocean*, Vol.6, pp.223-228 (in Japanese).

Chapter 8

Euler–Lagrange Hybrid Method

Hidemi Mutsuda

In addition to Eulerian surface capturing (the VOF method in Chapter 4), semi-Lagrangian modeling (the CIP method in Chapter 5) and the full Lagrangian particle method (Chapter 6), some Euler–Lagrange hybrid numerical schemes have been recently proposed. This hybrid method simultaneously supports both accurate definition of the free-surface/interface and efficient computation; Lagrangian particles arranged near the surface are tracked using a high-order finite difference scheme coupled with Eulerian fluid computations. Compared to Lagrangian methods that require many particles arranged over the entire fluid region, this coupled process can save in computing costs. One of these hybrid methods, a gas-liquid-solid three-phase computation proposed by Mutsuda,[15–18] is outlined in this chapter.

8.1. Introduction

Pioneering computational methods based on an Euler-Lagrange approach are the well-known Maker-and-Cell (MAC)[8] and Particle-in-Cell (PIC)[7] methods. In the MAC method, massless makers arranged over the fluid domain are passively transported with an advection velocity computed on Eulerian grids, which simply defines the surface by the maker locations. In the PIC method, physical quantities, including momentum and density, are computed on grids and interpolated to the particle locations; Lagrangian particles then directly transport these quantities through particle advection. For the next time step, the quantities from the particles are returned to the grids by interpolation, so the computing can be performed on the grids. Historically the PIC method has made significant contributions to research in computational fluid dynamics. The CIP method has been explained in Chapter 5, and the SPH method has been covered in Chapter 6; both of these methods were developed as an extension of the PIC approach.

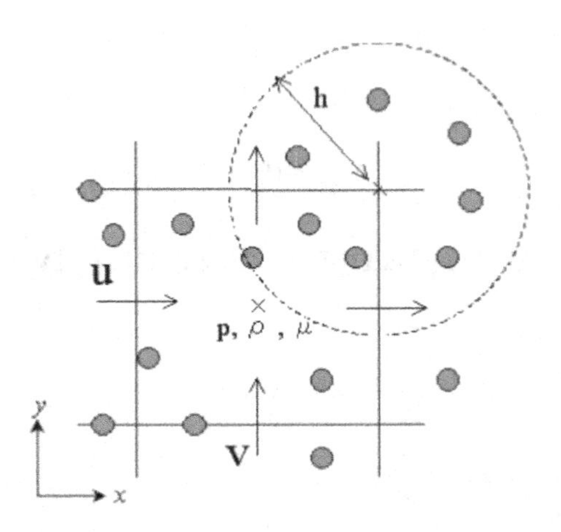

Fig. 8.1. Arrangement of grid and particles.

Other types of hybrid methods have also been developed recently. In the level-set framework, a particle level-set method[3] achieves accurate surface definition via Lagrangian advection of differently sized particles. Losasso *et al.*[14] proposed a two-way hybrid method composed of SPH and particle level-set approaches.

Other hybrid methods include: combinations of MPS and Eulerian finite difference models (Liu *et al.*[13]), MPS and CIP (Ishii *et al.*[9]) methods, and so on. Mutsuda *et al.*[15] developed another hybrid technique to compute three-phase gas-liquid-solid interactions based on CIPs, SPH, and Lagrangian particle tracking. The technique developed by Mutsuda *et al.* is outlined in §8.2, and the reader is referred to a series of references[15–18] for detailed interpretations of this method. Numerical validations and some applications to wave-structure interactions are introduced in §8.3.

8.2. Numerical Methods

A unified technique for compressible and incompressible fluids, the CCUP method (see §5.3.1) is applied to computations of air and water flows on Eulerian staggered grids (§3.2.1). Each phase, whether gas, liquid or solid, is identified by computing the advection equation for the density function (see §5.4.1):

$$\frac{\partial \rho_f^I}{\partial t} + \boldsymbol{u} \cdot \nabla \rho_f^I = 0 \tag{8.1}$$

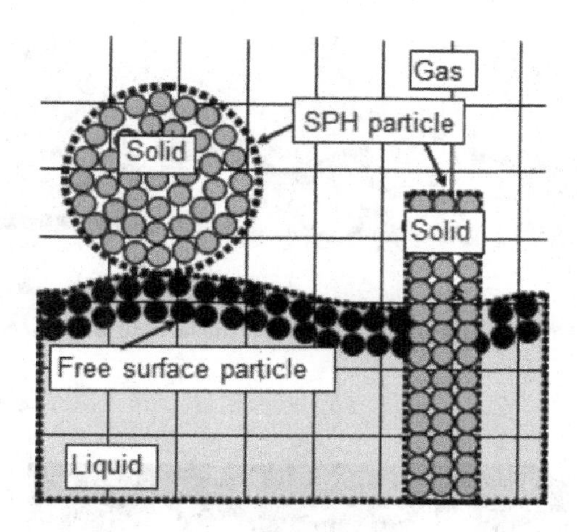

Fig. 8.2. Schematic representation of particle locations and Eulerian grids in a domain.

where I indicates the phase: gas = 1, liquid = 2, and solid = 3. Two kinds of particles are also introduced to distinctly identify displacements in the air-water interfaces and to correct the density function ρ_f^I in the solid body surfaces; this can reduce computational error that can occur in the surface capturing procedure. The free-surface, massless particles are arranged in a finite width on the liquid side of the interface, while the solid particles constitute a whole solid body or structure (see Figs. 8.1–8.4). The free-surface particles are updated through Lagrangian transport due to the local velocity computed in the Eulerian CCUP framework, and the deformation of the solid body is described by displacement of the solid particles that are computed with an SPH approach.

8.2.1. *Computation of free-surface particles*

The advection of free-surface particles is governed by the following ordinary differential equation.

$$\frac{d\boldsymbol{x}_p}{dt} = \boldsymbol{u}(\boldsymbol{x}_p) \tag{8.2}$$

where $\boldsymbol{x}_p$ is the particle location and $\boldsymbol{u}(\boldsymbol{x}_p)$ the particle velocity. The above equation is integrated using a fourth order accurate Runge-Kutta method. The variable $\boldsymbol{u}(\boldsymbol{x}_p)$ at the particle location is determined with a bilinear or trilinear interpolation of the Eulerian velocities computed on neighboring grids.

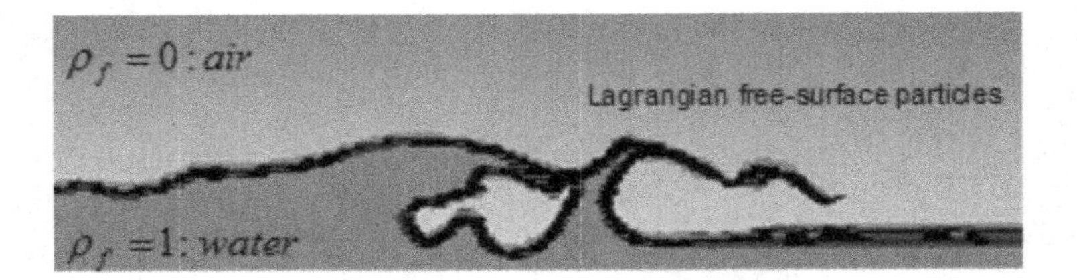

Fig. 8.3. Example of a particle distribution near the free-surface in a computed breaking wave.

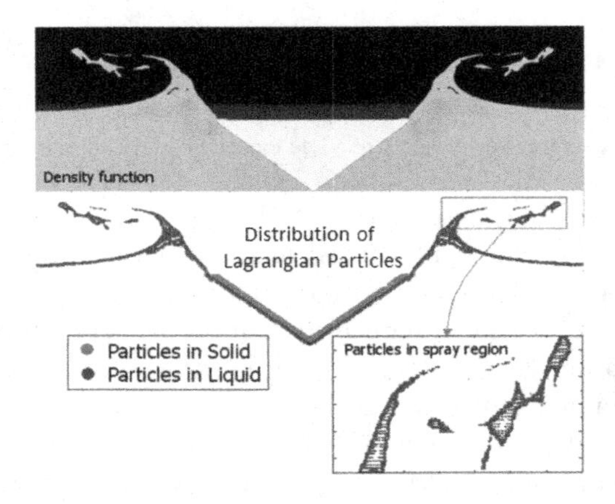

Fig. 8.4. Distributions of the density function (top) and free-surface/solid particles (bottom) around a splashing wedge.

During computation, the number density of the free-surface particles may locally change on the deformed free-surface. To maintain a smooth surface with a uniform number density, some particles should be removed from the dense part and added onto coarse surfaces to reduce possible numerical errors that occur on highly curved surfaces. To rearrange the particles with adjusted number densities, it is convenient to introduce a particle redistribution technique proposed by Enright[3] (see Fig. 8.5). This technique is based on the particle level set method, and the particle at location $\boldsymbol{x}_{p,old}$ is attracted toward a free-surface location as defined by the level set function $d = 0$ (see also §5.4.1); this is expressed as

$$\boldsymbol{x}_{p,new} = \boldsymbol{x}_{p,old} + \lambda \left(d_{goal} - d(\boldsymbol{x}_p) \right) \boldsymbol{n} \tag{8.3}$$

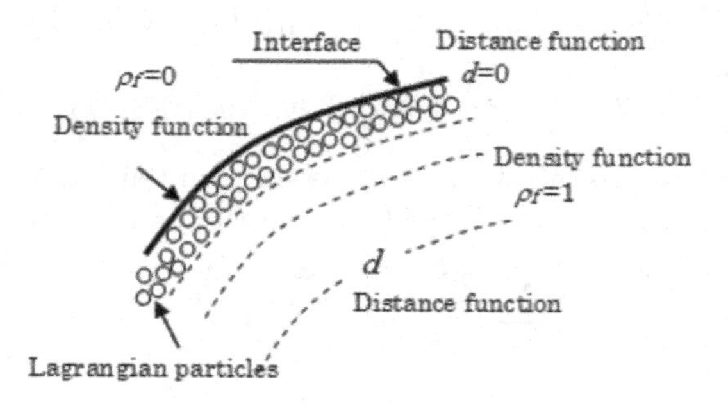

Fig. 8.5. Redistribution of free-surface particles near the free surface.

where $\boldsymbol{x}_{p,new}$ is the updated particle location, $\lambda = 1$, and the unit normal vector $\boldsymbol{n} = \nabla d/|\nabla d|$. The target distance from the surface d_{goal} should be empirically given to be suitable to the flow field.

8.2.2. *Computation of solid particles*

How the fluid forces deform the solid body or structure is directly computed in the SPH framework. This solution technique is presented below.

Conservation equations of mass and momentum for the solid elements are described by

$$\frac{d\rho}{dt} + \rho\frac{\partial u_i}{\partial x_i} = 0 \tag{8.4}$$

$$\frac{du_i}{dt} = \frac{1}{\rho}\frac{\partial \sigma_{ij}}{\partial x_j} + g_i + F_i \tag{8.5}$$

where ρ, u_i, σ_{ij}, and F_i represent the density, velocity, stress tensor, and fluid-solid interaction terms, respectively. The stress tensor is expressed as

$$\sigma_{ij} = \frac{\sigma_{kk}}{3}\delta_{ij} + S_{ij} \tag{8.6}$$

where S_{ij} is the deviation component of the stress tensor. The time increment for S_{ij} should be governed by the constitutive equation:

$$dS_{ij} = \mathbf{D}d\varepsilon_{ij} \tag{8.7}$$

where $\mathbf{D}$ is the elastic-plastic matrix and $d\varepsilon_{ij}$ is the time increment of the strain. The well-known Jaumann derivative, which ensures material frame indifference with respect to rotation, can be used to compute S_{ij}:

$$\frac{dS_{ij}}{dt} = 2\mu\left(\dot{\varepsilon}_{ij} - \frac{1}{3}\delta_{ij}\dot{\varepsilon}_{ij}\right) + S_{ik}\Omega_{jk} + \Omega_{ik}S_{kj} \quad (8.8)$$

where $\dot{\varepsilon}$ indicates the strain rate, and Ω_{ij} is the spin tensor.

The SPH representations[5] of Eqs. (8.4) and (8.5) are written as

$$\frac{d\rho^n}{dt} = \rho^n \sum_l \frac{m^l}{\rho^l}\left(u_i^n - u_i^l\right)\frac{\partial W^{nl}}{\partial x_i^n} = 0 \quad (8.9)$$

$$\frac{du_i^n}{dt} = \sum_l m^l \left(\frac{\partial \sigma_{ij}^n}{(\rho^n)^2} + \frac{\partial \sigma_{ij}^l}{(\rho^l)^2} + \Pi^{nl}\delta_{ij} + R_{ij}^{nl}\right) + g_i - \frac{F_i^n}{\rho^n} \quad (8.10)$$

where m^n is the mass of the n-th particle, ρ^n is the density, W^{nl} is the kernel function (a third-order Spline function), and Π^{nl} is the artificial viscosity defined by Gingold *et al.*[5] R_{ij}^{nl} is the artificial stress to remove the tensile instability (see Gray *et al.*[6] for details). The interaction term F^n is estimated from the fluid pressure and interpolated to the solid particles:

$$F^n = -\frac{1}{\rho^n}\sum_l m^l \frac{p^l}{\rho^l}\nabla^n W^{nl} \quad (8.11)$$

For a case where the solid phase can be assumed to be a rigid body, it is convenient to merely estimate the translation and rotation of the centroid to describe the motion of the object instead of conducting the time-consuming computation for the individual solid particles. In this instance, the governing equations for rigid body motion are expressed by

$$\frac{\partial^2 \boldsymbol{r}_g}{\partial t^2} = \frac{\boldsymbol{G}}{M} - \boldsymbol{F} \quad (8.12)$$

$$I\frac{\partial \boldsymbol{\omega}}{\partial t} = \boldsymbol{T} \quad (8.13)$$

$$\frac{\partial \boldsymbol{\theta}}{\partial t} = \boldsymbol{\omega} \quad (8.14)$$

where $\boldsymbol{r}_g$ is the centroid of body, $\boldsymbol{G}$ is the total external force acting on the rigid body, $\boldsymbol{F}$ is the fluid-solid interaction, M is the total mass of the rigid body (estimated as the sum of all particle masses that compose the rigid body), I is the inertia tensor, $\boldsymbol{\omega}$ is the angular velocity, $\boldsymbol{T}$ is the torque, and $\boldsymbol{\theta}$ is the rotation angle.

The parameters $\boldsymbol{r}_g$ and I are estimated as[2]

$$\boldsymbol{r}_g = \frac{1}{N}\sum_{n=1}^{N} \boldsymbol{r}_n \quad (8.15)$$

$$I = \sum_{n=1}^{N} m_n |\boldsymbol{r}_n - \boldsymbol{r}_g|^2 \tag{8.16}$$

where $\boldsymbol{r}_n$ is the n-th particle location, N is the total number of particles constituting the body, and m_n is the mass of the n-th particle.

The particle velocity $\boldsymbol{u}_n^{k+1}$ and location $\boldsymbol{r}_n^{k+1}$ at time step $k+1$ are updated with the following procedure.

$$\hat{\boldsymbol{r}_n}' = \hat{\boldsymbol{r}_n}^{k+1} - \boldsymbol{r}_n^k \tag{8.17}$$

$$\boldsymbol{r}_g^{k+1} = \frac{1}{N}\sum_{n=1}^{N} \hat{\boldsymbol{r}_n}^{k+1} \tag{8.18}$$

$$\boldsymbol{r}_g' = \frac{1}{N}\sum_{n=1}^{N} \hat{\boldsymbol{r}_n}' \tag{8.19}$$

$$\boldsymbol{\theta}' = \boldsymbol{\omega}\Delta t = \frac{1}{N}\sum_{n=1}^{N} m_n \hat{\boldsymbol{r}_n}' \times \left(\boldsymbol{r}_n^k - \boldsymbol{r}_g^{k+1}\right) \tag{8.20}$$

$$\boldsymbol{r}_n' = \boldsymbol{r}_g' + \boldsymbol{R}^{-1}\left(\boldsymbol{r}_n^k - \boldsymbol{r}_g^{k+1}\right) \tag{8.21}$$

$$\boldsymbol{u}_n^{k+1} = \frac{\boldsymbol{r}_n'}{\Delta t} \tag{8.22}$$

$$\boldsymbol{r}_n^{k+1} = \boldsymbol{r}_n^k + \boldsymbol{r}_n' \tag{8.23}$$

where R is the 3×3 rotation matrix, and Δt is the time-step interval.

While there are several major descriptions of rotation in three-dimensional space, following Baraff,[2] R is determined on the basis of representation by the unit quaternion, $\boldsymbol{q}(t) = s + v_x i + v_y j + v_z k$, where i,j, and k are the respective imaginary units, s is a scalar value, and $\boldsymbol{V} = (v_x, v_y, v_z)$ is the unit vector of the rotational axis.

$$\boldsymbol{R} = \begin{pmatrix} 1 - 2v_y^2 - 2v_z^2 & 2v_x v_y - 2sv_z & 2v_x v_z + 2sv_y \\ 2v_x v_y + 2sv_z & 1 - 2v_x^2 - 2v_z^2 & 2v_y v_z - 2sv_x \\ 2v_x v_z - 2sv_y & 2v_y v_z + 2sv_x & 1 - 2v_x^2 - 2v_y^2 \end{pmatrix} \tag{8.24}$$

The reader should refer to §7.2.1 for the details of this approach.

8.3. Applications of the Hybrid Model

8.3.1. *Numerical test*

The ability of capturing surfaces is examined with stretched flows, a so-called vortex-in-vortex problem. The liquid phase is initially defined as a

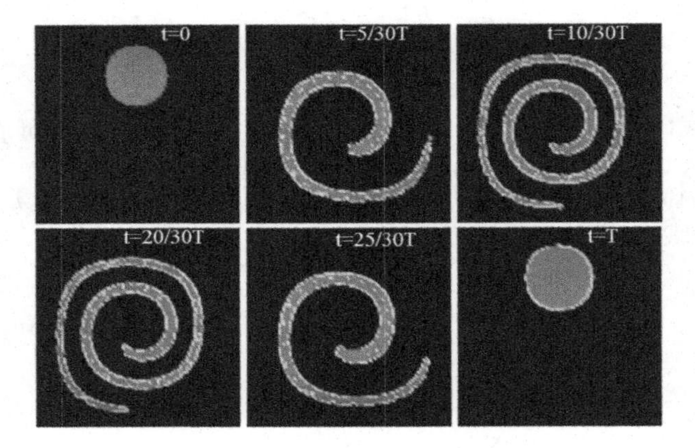

Fig. 8.6. Deformation of a liquid patch in a stretched flow (four particles per cell are assigned near the interface).

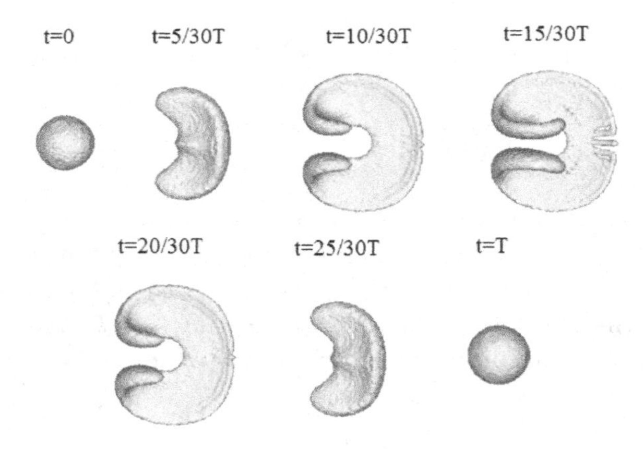

Fig. 8.7. Deformation of a liquid patch in a three-dimensional stretched flow (eight particles per cell are assigned near the interface).

circle of radius 0.15 placed at (0.5, 0.75) and surrounded by a gas phase in a unit computational domain. The velocity of the two-dimensional stretching flow is given by

$$u(x, y) = -2\sin^2(\pi x)\sin(\pi y)\cos(\pi y)\cos(\pi t/T) \tag{8.25}$$

$$v(x, y) = 2\sin^2(\pi y)\sin(\pi x)\cos(\pi x)\cos(\pi t/T) \tag{8.26}$$

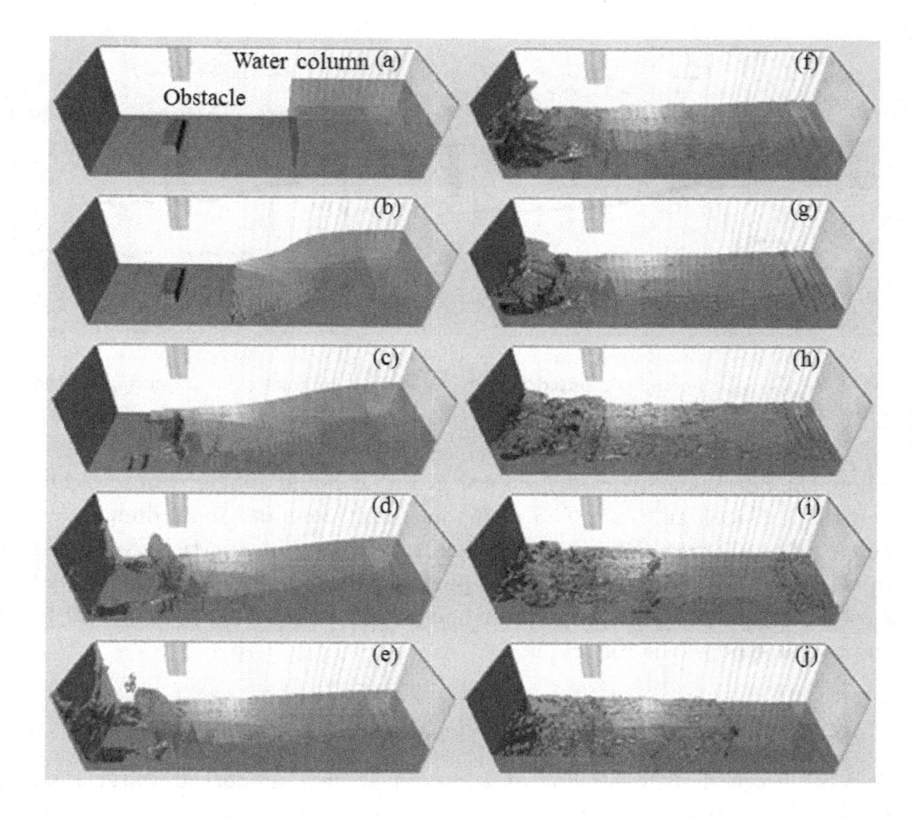

Fig. 8.8. Computed surface evolution of the dam-break flow in a channel with a rectangular structure composed of solid particles.

where the normalized period T=10 was given. To capture the interface, 1,200 free-surface particles were used in a 100×100 grid cell domain for this test. This vortical flow stretches the liquid circle into a long filamentary pattern (see Fig. 8.6). The thin filament is well resolved, and after one period ($t = T$), the initial circular pattern of the liquid phase is finally restored.

Next, the computed three-dimensional deformation of an interface is examined in the three-dimensional stretched flow with the following velocity:

$$u(x, y, z) = 2\sin^2(\pi x)\sin(2\pi y)\sin(2\pi z)\cos(\pi t/T) \tag{8.27}$$

$$v(x, y, z) = -\sin(2\pi x)\sin^2(\pi y)\sin(2\pi z)\cos(\pi t/T) \tag{8.28}$$

$$w(x, y, z) = -\sin(2\pi x)\sin(2\pi y)\sin^2(\pi z)\cos(\pi t/T) \tag{8.29}$$

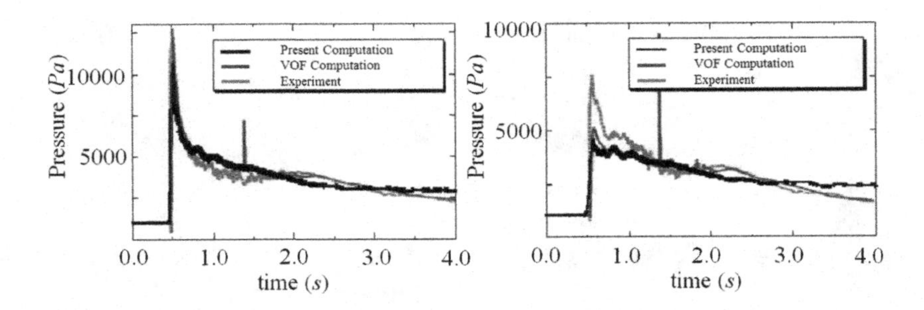

Fig. 8.9. Comparing the computed time records of the impact pressure acting on the structure faces with experimental results.[10]

where the normalized period T=3 was given. A liquid sphere, whose radius is 0.15, is placed at (0.35, 0.35, 0.35) in a unit computational domain. A $100 \times 100 \times 100$ grid cell domain with 45,123 free-surface particles was used in this test. Although the interface partially disappears in under-resolved regions, the final liquid shape after one period is found to be identical to the initial sphere (see Fig. 8.7).

8.3.2. *Interfacial flows with solid structures*

A well-known dam-break flow is reproduced in a numerical water tank (Fig. 8.8). After the progressing bore hits a rectangular structure installed in the middle of the channel, an uprushing jet is formed on the face of the structure, while a partial bore passes beside the structure and hits the end wall of the channel, being reflected. Figure 8.9 shows graphs to compare the computed time records of the impact pressure acting on the structure faces with the experimental results of Kleefsman *et al.*;[10] the computational results show consistent agreement with the experiment.

Next, the computed results of breaking waves propagating over a reef, which hit an end wall in a numerical wave flume, are shown in Fig. 8.10. The complex shapes of the interface that forms during the violent splash-up process is reasonably computed. In this computation, the number of grids, whose widths are 2.68 mm in the x-direction and 3.89 mm in the y-direction, is 248 57 in a planar two-dimensional domain. The time step interval is 10^{-4}s. A total of 4,700 particles capture the free-surface and are initially located at the free surface. The volume error (total mass loss) was less than 0.05% throughout the computation. Figure 8.11 shows the computed

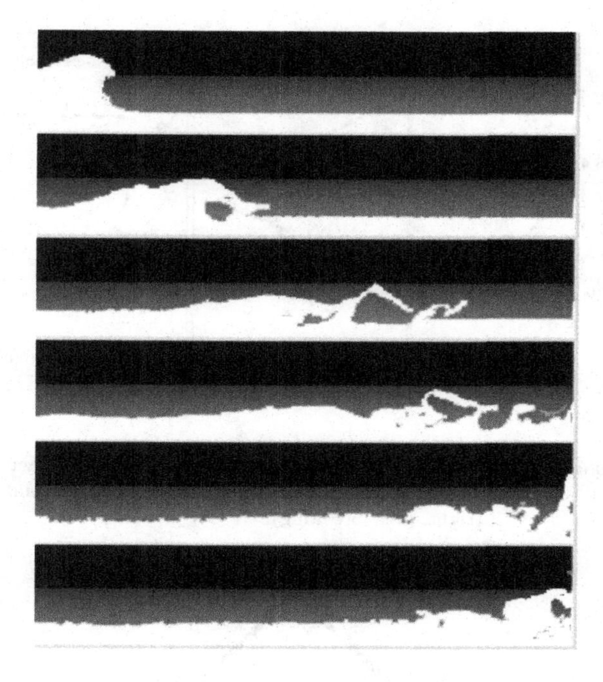

Fig. 8.10. Evolution of waves breaking on a reef in a numerical wave tank with an end wall.

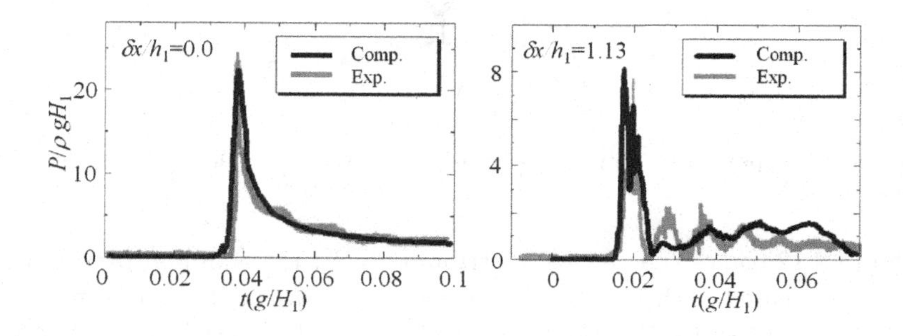

Fig. 8.11. Time histories of the impact pressures acting on the wall.

time records of dynamic pressure acting on the end wall when the breaking wave hits it, as they compare with the experimental values from Azarmsa *et al.*[1] The overlay plots indicate that the overall features of the evolution of the computed pressures are identical with the experimental values.

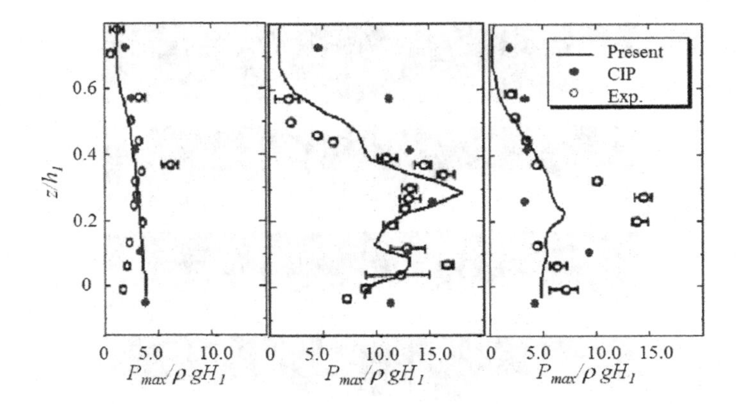

Fig. 8.12. Comparing the vertical distributions of the maximum impact pressures on the vertical wall at δ_x/h_1 = -0.53 (left, pre-breaking), δ_x/h_1 = 0.47 (middle, breaking point), and δ_x/h_1 = 1.13 (right, post-breaking).

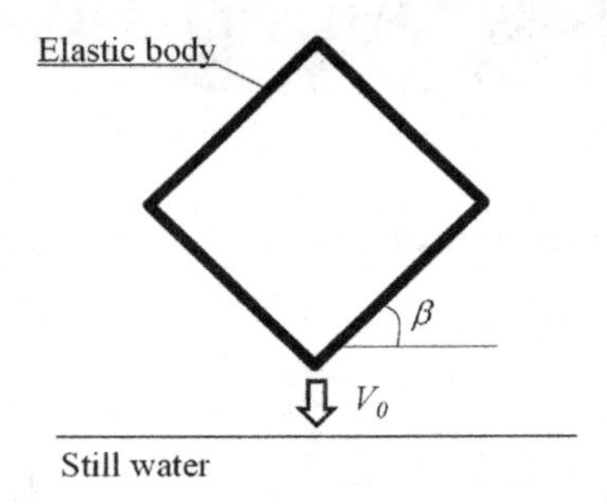

Fig. 8.13. Experimental setup for a splashing elastic body (V_0 : entry velocity, β : inclined angle).

Figure 8.12 shows the vertical distributions of the maximum pressures on the wall located at three horizontal positions: δ_x/h_1 = -0.53 (pre-breaking), δ_x/h_1 = 0.47 (breaking point), and δ_x/h_1 = 1.13 (post-breaking), where δ_x is the horizontal distance from the breaking point, and h_1 is the water depth. Overall the maximum pressure values agree with experimental results, indicating that computational methods allow for reliable estimates of the wave forces required to design coastal structures.

A water entry test of an elastic body was performed to examine the hydro-elastic responses computed with the SPH elastic model (Eqs. (8.8)–(8.11)) as compared with experimental ones. A rectangular elastic box was

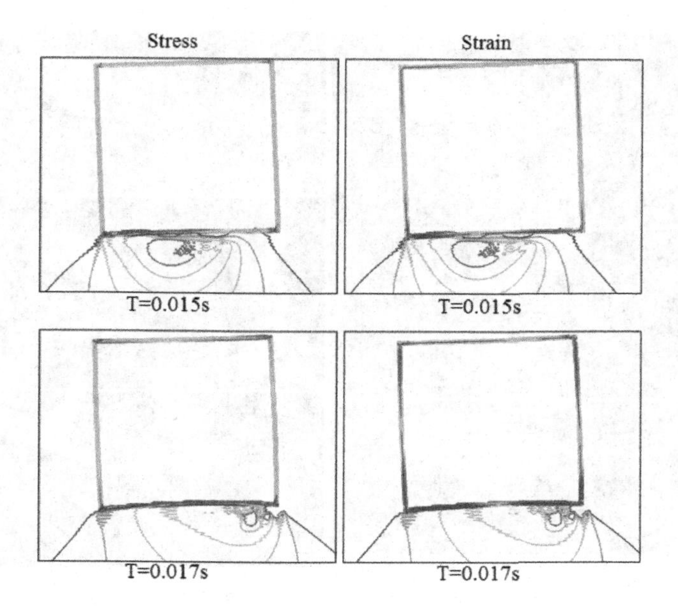

Fig. 8.14. Contour distributions of internal stress and strain in the elastic body, and water pressure during the water entry process ($\beta = 2.5^o$).

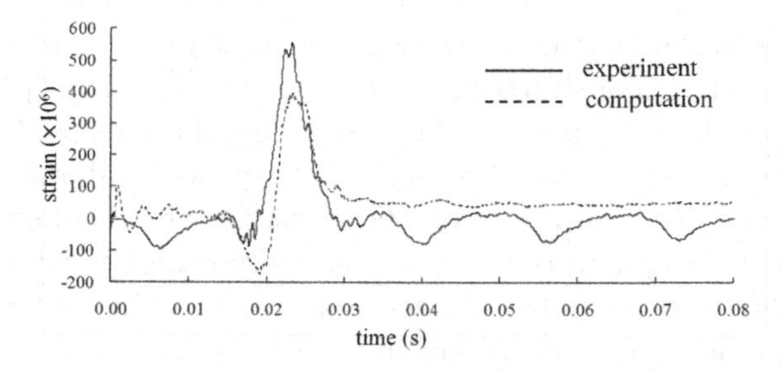

Fig. 8.15. Comparing the computed time history of the strain with the experimental history ($\beta = 15^o$).

dropped into still water with an inclined angle β relative to the horizontal axis (see Fig. 8.13). The pressure, stress and strain were measured with piezo-electric sensors installed on the bottom of the elastic body. The elastic rectangular body contained 3,900 solid particles, each with a radius of 0.25 mm to compute pressure, stress, and strain at the bottom surface

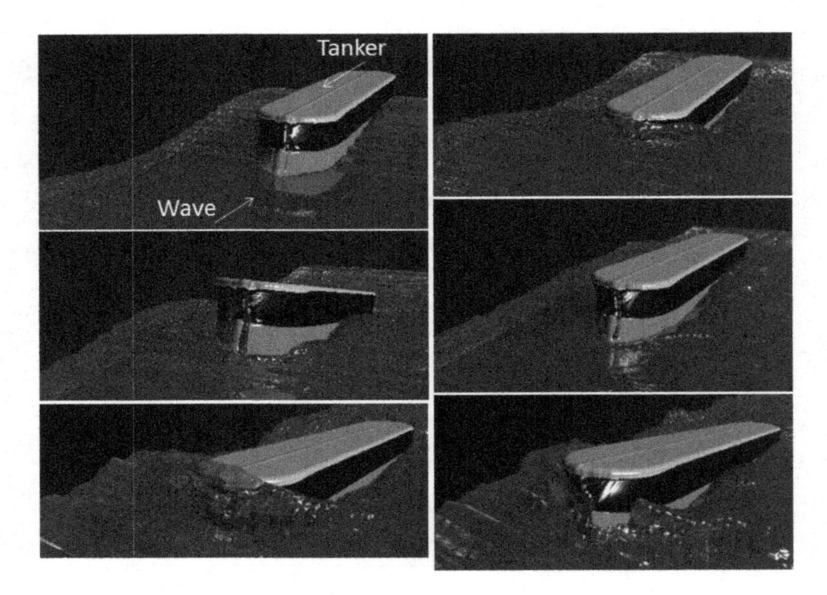

Fig. 8.16. Three-dimensional behaviors of a tanker responding to a head sea of waves.

in a high resolution. The size of the Eulerian rectangular grid was 2 mm, and the total number of free-surface particles, whose radii also measured 0.25 mm, was 6,024. Two different time-step intervals were used in this multi-phase computation due to the different mechanical responses of the fluid and solid: an interval of 5.0×10^{-4} s for the liquid and gas phases, and 2.5×10^{-5} s for the solid phase. The spatial distributions of the internal strains on the elastic body, and the impact water pressures during the water entry process are shown in Fig. 8.14. The internal stress and strain occur locally within the wall, and the bottom especially is highly deformed after splashing. Also, the computed impact water pressure is intensified near the bottom where strong internal strain values appear. Figure 8.15 shows the time record of strain, comparing the computational values with the experimental results; these comparisons also demonstrate how this type of multi-phase computation is appropriate.

Finally, we investigate the computed behaviors of a tanker (as a floating elastic body) in a wave field (see Fig. 8.16). The complex heaving and pitching motions of a tanker besieged by a head sea of high waves are reasonably computed, along with the splashing, breaking, and slamming events of the waves against the tanker. In Fig. 8.17, the heave and pitch

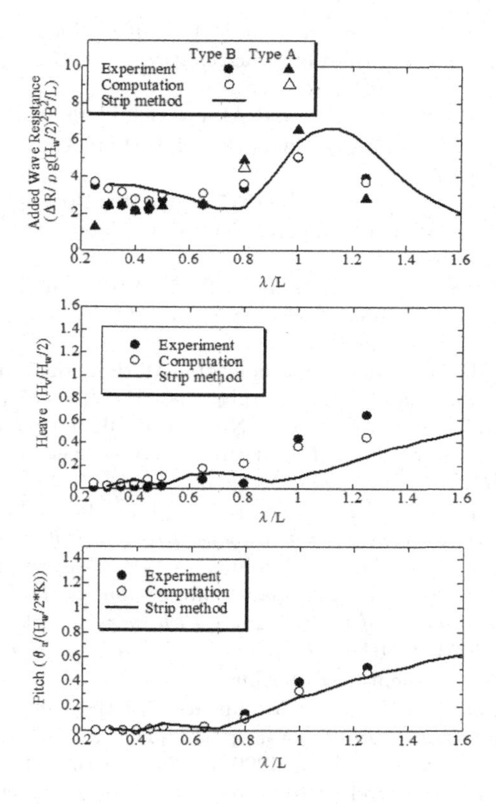

Fig. 8.17. Comparisons of the added resistance (top), heaving (middle), and pitching (bottom) motions between numerical and experimental results.

of the tanker and the added wave resistance are compared with both experimental results and calculations based on strip theory.[11] The plots in Fig. 8.17 indicate that the numerical computations are consistent with the experimental results, providing much better predictions than strip theory.

References

1. Azarmsa S.A. (1996): *Impact pressure and decay properties of breaking waves*, Dr thesis, Gifu University.
2. Baraff D. (1997): An introduction to physically based modelling : Rigid body simulation I 'Unconstrained rigid body dynamics', *SIGGRAPH'97 course notes*, D3.

3. Enright D., Fedkiw R., Ferzger J. and Mitchell I. (2002): A hybrid particle level set method for improved interface capturing, *J. Computational Physics*, Vol.183, pp.83-116.
4. Ghosal S., Lund T.S., Moin P. and Akselvoll K. (1995): A Dynamic Localization Model for Large Eddy Simulation of Turbulent Flows, *J. Fluid Mech.*, Vol.289, pp.229-255.
5. Gingold R.A. and Monaghan J.J. (1977): Smoothed particle hydrodynamics, theory and application to non-spherical stars, *Mon. Not. Roy. Astr. Soc.*. Vol.181, pp.375-389.
6. Gray J.P., Monaghan J.J. and Swift R.P. (2001): SPH elastic dynamics, *Computer methods in Applied Mechanics and Engineering*, Vol.190, pp.6641-6662.
7. Harlow F.H. (1955): A Machine Calculation Method for Hydrodynamic Problems. *Los Alamos Scientific Laboratory report LAMS-1956.*
8. Harlow F.H. and Welch J.E. (1965): Numerical calculation of time-dependent viscous incompressible flow of fluid with a free surface. *Phys. Fluids*, Vol.8, pp.2182-2189.
9. Ishii E., Ishikawa T. and Tanabe Y. (2004): Particle/CIP hybrid method for predicting motions of micro free surfaces, *Proc. ASME* HT-FED2004-56142.
10. Kleefsman K.M.T., Fekken G., Veldman A.E.P., Wanowski B. and Buchner B. (2005): A Volume-of-Fluid based simulation method for wave impact problems, *Int. Journal of Computational Physics*, Vol.206, pp.363-393.
11. Lewandowski E.M. (2004): *The dynamics of marine craft maneuvering and seakeeping*, World Scientific Publishing.
12. Lilly D.K. (1992): A Proposed Modification of the Germano Subgrid-Scale Closure Method, *Phys. Fluids*, A4, pp.633-635.
13. Liu L., Koshizuka S. and Oka Y. (2003): Computation of multiphase flow by coupling the MPS method with mesh method, Spring meeting 2003, *Atomic Energy Society of Japan*, pp.479.
14. Losasso F., Talton J., Kwatra N. and Fedkiw R. (2008): Two-way coupled SPH and particle level set fluid simulation, *IEEE TVCG*, Vol.14, pp.797-804.
15. Mutsuda H. and Doi Y. (2008): Highly accurate Free Surface Capturing Technique for Wave Breaking, *Proc. Int. 31st Conf. on Coastal Engineering*, pp.38-50.
16. Mutsuda H., Shimizu Y. and Doi Y. (2009): Numerical Study on Interaction between Violent Wave and Structure using SPH, Particle-Based Methods, -Fundamentals and Applications, *PARTICLES 2009*, pp.266-269.
17. Mutsuda H. and Doi Y. (2009): Numerical Simulation of Dynamic Response of Structure Caused By Wave Impact Pressure Using an Eulerian Scheme with Lagrangian Particles, *Proc. ASME 28th Int. Conf. on Ocean*, Offshore and Arctic Engineering, OMAE2009-79736, CD-ROM.
18. Mutsuda H., Kurokawa T., Suandar B. and Doi Y. (2010): Numerical Simulation of Interaction Between Wave and Floating Body using Eulerian Scheme with Lagrangian Particles, *IV European Conf. on Computational Mechanics*, CD-ROM.
19. Osher S. and Fedkiw R.: *Level set methods and dynamic implicit surfaces*, Springer-Verlag.

20. Suandar B., Mutsuda H. , Kurihara T., Kurokawa T., Doi Y. and Shi J.(2011): An Eulerian Scheme with Lagrangian Particle for Evaluation of Seakeeping Performance of Ship in Nonlinear Wave, *Int. J. Offshore and Polar Engineering*, Vol.21(2), pp.103-110.
21. Suandar B., Mutsuda T. and Doi Y. (2011): Numerical Study on Propulsion and Seakeeping Performance of a Ship Using an Eulerian Scheme with Lagrangian ParticlesC*Journal of the Japan Society of Naval Architects and Ocean Engineers*, Vol.13, pp.19-26.

Chapter 9

Computational Wave Dynamics for Coastal and Ocean Research

Yasunori Watanabe, Nobuhito Mori, Hitoshi Gotoh and Akio Okayasu

Computational wave dynamics (CWD) is a specialized branch of fluid mechanics to analyze and numerically solve flows under water waves using computational techniques. What distinguishes CWD is the focus on the free-surface of water waves, computing the interfacial interactions, and describing the processes that occur. CWD is a specialized subset of computational fluid dynamics (CFD). A numerical wave flume (NWF) is defined to be a computational simulator of coastal and ocean waves using a CWD model*.

In this chapter, the current and future prospects of NWF research and applications will be discussed, as they pertain to coastal issues involving sediment transport (§9.1), wave-structure interaction (§9.2), and coastal disasters (§9.3). We also briefly discuss possible avenues of future research within the field of CWD.

9.1. Sediment Transport

The surf zone is the boundary between land and ocean, where material transport is significantly enhanced due to the complex unsteady flows produced during the wave breaking process. In terms of material transport, the surf zone plays a role at sourcing and distributing sediment materials along the shoreline and inshore areas of an ocean (see Fig. 9.1). For instance, unsteady organized flows produced by breaking waves govern suspension processes of fine sediments on a beach, and therefore a quantitative estimate of the suspended sediment concentration inside the surf zone is essential to achieve reliable predictions of turbidity distributions and sediment transport on a regional scale.

*A numerical wave tank (NWT) has been commonly used to express a wave simulator in which a boundary integral method or depth-integrated wave equation is originally used to compute surface elevation. To distinguish the methodology and governing equations from those used in a NWT, we will use NWF in this book rather than this definition of NWT.

Fig. 9.1. Typical beach scarp formed at Mukawa beach (Hokkaido, Japan).

Modifications of the local wave field, e.g., due to artificial wave control, may disrupt the equilibrium state of coastal bathymetry. In practical shoreline control, shoreline displacement may be interpreted on the basis of past records of regional sediment budgets, regardless of any mechanical transport process. However, the reciprocal responses occurring with the local wave transformation, wave-induced nearshore currents, and the three-dimensional changes in bathymetry that result from local sand drift, need to be understood; this allows for long-term assessment of a prospective equilibrium state of coastline geometry and seabed form, as well as the local wave and current fields that depend on them. The modified local distributions of fine suspended sediments also influence the habitats of marine plants and animals, water quality and insolation in the sea, which may be an issue for coastal marine environments and the fishery industry.

Conventional estimates of sediment suspension due to wave motion were predicted using three characteristic parameters: the fluid velocity, the

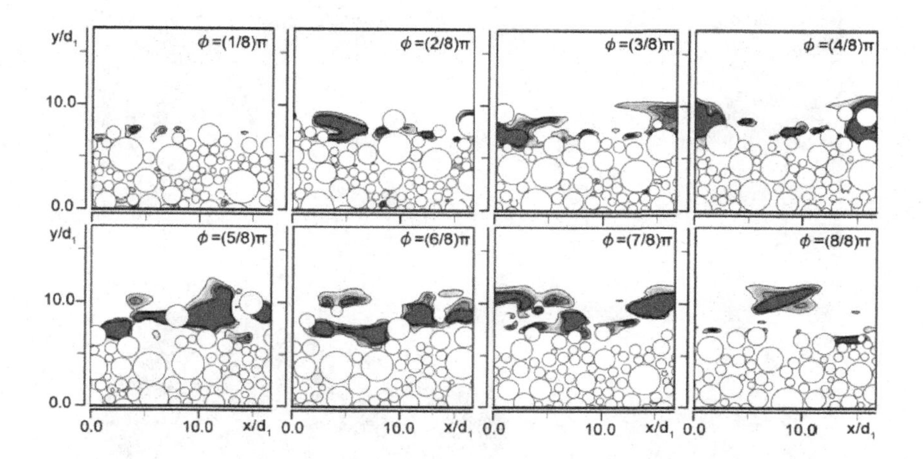

Fig. 9.2. Evolution of turbulent energy in the wave boundary layer over non-cohesive sands;[6] ϕ: wave phase.

diffusion coefficient (passiveness of sediment behavior with respect to fluid flow), and a reference concentration (bottom boundary condition for the sediment concentration). The prediction was based on the one-dimensional (vertical) diffusion equation solved under a quasi-steady state assumption. Appropriately assessing the values of these parameters is difficult, and they were conventionally estimated with empirical relationships for the surf zone where material and momentum are consecutively exchanged. Two parameters are difficult to estimate: the superficial diffusion coefficient and the local velocity, especially in the turbulent, complex flow field that occurs during wave splashing. Because turbulence on the beach bottom causes strong disturbances that suspend the local sediments, the rational reference level of the bottom surface may be undefined (see Fig. 9.2), and the simple diffusion model itself may not be justified in this case. The flow structures that govern sediment suspensions need to be reasonably predicted to improve the turbidity estimates in the complex carrier flow, a prediction that a NWF is expected to achieve (see Fig. 9.2).

In the past, the most difficult component of wave-breaking computations was the numerical definition of complex-shaped free-surfaces. As described in the previous chapters, many free-surface tracking and detecting techniques, including particle methods (Chapter 6), VOF (Chapter 4), the density function and level-set methods (Chapter 5), have been developed

Fig. 9.3. Breaking waves hitting a rubble structure.

to achieve precise and reliable surface definitions. These techniques have been applied to breaking waves and led to some important findings for breaking dynamics, as already described. Some pilot applications of the multi-phase flow computation of the surf zone with mixtures of air-bubbles and suspended sediments have also been undertaken (see §5.5 and §8.3), and these applications are on the cusp of being extended to CWD and practical wave-breaking research.

Recently many easily-performable, open-source CFD codes with pre- and post-processing functions have been provided via Internet. While using readily accessible code may enhance motivation among students and young researchers of CFD, we hope they will also acquire a deep understanding of the mathematical models in CWD and become future experts who assume the role of coding developments for coastal and ocean research. We also hope this book will provide help to understand CFD and CWD techniques.

9.2. Wave-Structure Interaction

In coastal engineering, a NWF is practically used to estimate wave forces acting on coastal structures to determine the optimal design. Further applications of a NWF include diverse problems of wave deformation, wave control, material diffusion and transport, sediment processes (coastal erosion and sediment drift), and sea surface processes (momentum, heat, and gas exchanges). In this section, we discuss practical applications of CWD and potential capabilities of a NWF as applied to engineering design of coastal structures.

While a conventional, functional design of a coastal structure is based on empirical and deterministic concepts,[4] a recent reliability design[5] is based on probability. These recent designs use probability variables for diverse structural and mechanical factors that are evaluated to maintain the structure function during its life time, i.e., errors in estimating the external force, randomness of the external force, incompleteness and uncertainty of the hydraulic data, and uncertainty in material tests. These factors were not able to be sufficiently incorporated in previous empirical analyses.

There are specific design formulas corresponding to specific types of structures, such as a pile structure, breakwater, and rubble structure. The well known Morison formula[13] has been used to estimate the wave force acting on a pile structure; the equation is semi-empirical and composed of an inertia force in phase with the local flow acceleration dU/dt and a drag force proportional to the square of the flow velocity $U(t)$:

$$F = \rho C_M V \frac{dU}{dt} + \frac{1}{2}\rho C_D A U|U| \tag{9.1}$$

where V is the volume of the body, A is the cross-sectional area of the body perpendicular to the flow direction, and C_M and C_D are the inertia and drag coefficients, respectively. The coefficients used in practical problems were empirically determined from the measured fluid force acting on a single cylindrical body in a large flow domain assumed to be of infinite space (*i.e.*, no other adjacent bodies). However, since pile structures are often formed in group arrangements of multiple piles, how the flow interactions between the adjacent piles affects the coefficients should be taken into account. Although empirical corrections for specific pile arrangements under standard types of irregular waves may be possible, it is very difficult to generalize the influences for every pile arrangement under arbitrary wave conditions using only past experimental wave results. If wave pressures resulting from local flow around the piles can be properly computed in a NWF, the wave force acting on a pile group of any arrangement can be estimated without the use of empirical coefficients that apply to a particular experimental setup.

The mechanical stability of a vertical breakwater for sliding and overturning failure modes against wave forces is ensured in the current design

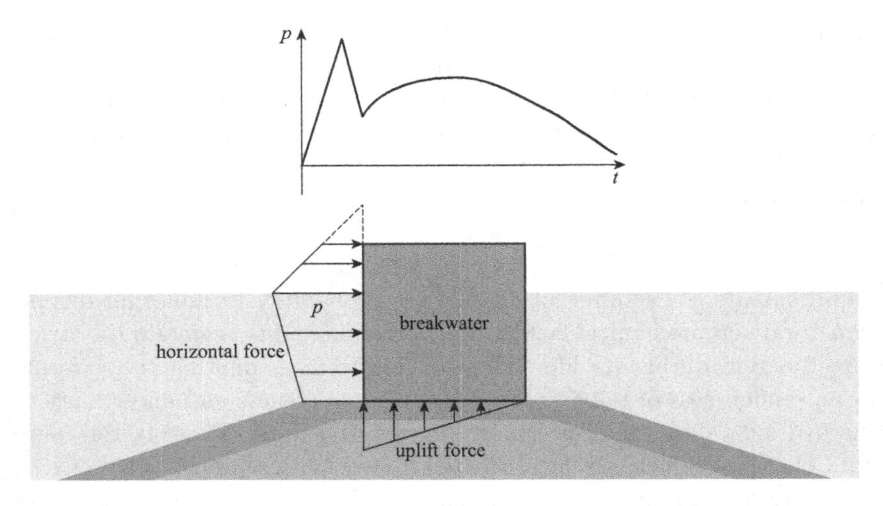

Fig. 9.4. Schematic representation of wave pressure on a breakwater (bottom) and its time series (top).

method, which is described with the well-known Goda formula[4] that accounts for both the wave pressure due to standing waves at the face of wall and the impact pressure from breaking waves (see Fig. 9.4). While this formula gives a simple estimate of the maximum pressure from design waves on a standard type of breakwater, additional empirical analysis of varying structural shapes and types is needed to improve the formula. Also, when additional causes induce another mode of failure, careful examinations may be needed. Local erosion and fluidization of the caisson mound and foundation affect the total stability of the structure, and the temporal variations of the pressure and local flow fields should be reasonably estimated (see Fig. 9.5). A NWF is expected to simultaneously compute these complex interactions between the wave, structure, and sediment in a framework of multiphase computations (see §4.4, §5.5, §6.6 and §8.3).

A composite breakwater covered with wave-dissipating armor blocks is the most popular type of breakwater in Japan. With this type of structure, displacement and subsidence of the armor blocks is another issue that needs to be assessed in the design, because subsidence reduces the covered section of the breakwater and intensifies the wave force acting on the caisson body. Also, the arrangement of the wave dissipating blocks in front of the breakwater is often important since wave pressures may be significantly intensified depending on the relative relationship between local wave

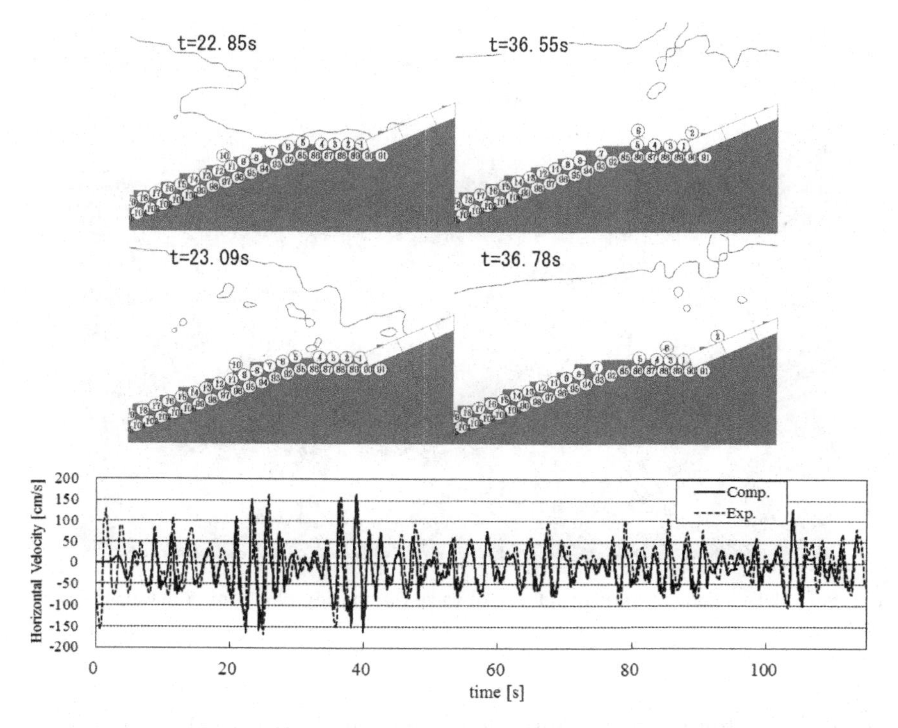

Fig. 9.5. Computed displacement of armor blocks arranged on a gently sloping seawall exposed to waves (top), and time record of local horizontal velocity near the armor layer[9] (bottom).

transformation and the envelope shape of the covered section. As mentioned in §7.2.2, a NWF can realistically and precisely estimate the behavior of the armor blocks and the body of the breakwater under local fluid forces; this capability potentially changes the current design methods for coastal structures.

The Hudson formula[7] has been used to determine the stable weight of armor units on rubble structures based on the static mechanical balance between the wave induced lift force (proportional to the square of the flow velocity) and gravity. The lift coefficient used in this formula was assumed to be identical to an empirical constant experimentally determined under flow conditions around a single cylinder in a relatively large domain (i.e., no interaction with an adjacent body). However, we find that in practice, using this constant may violate this assumption. In complex unsteady flows over

Fig. 9.6. Displacement and subsidence of wave-dissipating blocks on a detached breakwater.

the spaces between armor units, fluid forces under waves that run-up the structure produce shielding effects due to the nearby armor units, and this effect dominates in the armor unit layer on rubble structures (see Fig. 9.6). The Distinct Element Method coupled with a CFD model (see Fig. 9.5 and also §7.2.2) reasonably predicts the fluid forces.

A number of laboratory experiments have been performed to estimate run-up wave heights and wave overtopping rates for standard types of seawalls and breakwaters, which were parameterized by wave conditions and local bathymetry for use in functional designs. However, the diverse range in the experimental results is significantly large when compared with the mean quantities, especially when wave breaking events are involved. An NWF can quantitatively compute instantaneous run-up and overtopping processes and precisely define the free-surfaces during those events.

The above examples involve indeterministic processes with uncertain variations that result from strong nonlinear fluid dynamics, especially with wave breaking events that clearly indicate a statistical limitation (or uncertainty) with the conventional empirical design formula. Some integrated shore protection systems composed of multiple wave-control and sediment-control structures have been constructed in this decade. With conventional design formulas, multiple individual assessments within this complex system may be insufficient due to the multiple uncertain effects that are coupled in the system. In the past, a laboratory experiment was the only way to examine this kind of theoretically undetermined problem. And physical model experiments have an additional crucial problem that scale similarity may not be guaranteed. Therefore, an NWF is expected to be a powerful tool to efficiently and optimally design coastal structures in the near future.

9.3. Coastal Disasters, Wave Climates, and Ocean Modeling

Some of the major external forces that cause a coastal disaster are extreme waves due to tropical, extra-tropical and seasonal storms. There are several reasons why storms result in coastal disasters: unexpectedly large waves are generated offshore, unexpected local waves and fluid behavior, and unexpectedly large waves at the shore.

The first step to estimate the coastal wave force is analysis of the offshore wave height using long-term statistics based on extreme value analysis. An offshore design wave height is generally estimated with 20–50 years of measured or hindcasted wave data at a target location. The second step involves a wave transformation, including diffraction and refraction due to bathymetry changes in the target area. The wave transformation uses numerical models based on the energy balance equation or Boussinesq-type nonlinear wave equation. The final step at estimating coastal wave forces involves numerical analysis using a three dimensional numerical scheme, which was discussed in previous sections. A brief overview of the role of NWF in coastal and ocean research will be described in this section.

An offshore wave condition significantly depends on the season and the ocean basin. For example, a tropical cyclone produces severe extreme waves, but the duration is relatively short. On the other hand, a seasonal storm is relatively lower in intensity, but the duration is longer than a tropical cyclone. Regarding wave generation and propagation, a longer fetch generates larger waves and causes coastal disasters in the middle to upper latitudes. Fig. 9.7 shows the time history of significant wave heights and periods due to an extreme winter storm in the Sea of Japan. The maximum significant wave height was about 13 m, and wave heights over 5 m continued for more than 24 hours during this storm event. Considering the maximum wave forces and fluid behavior, the peak condition of this type of storm is selected for a detailed analysis. The measured offshore wave profile

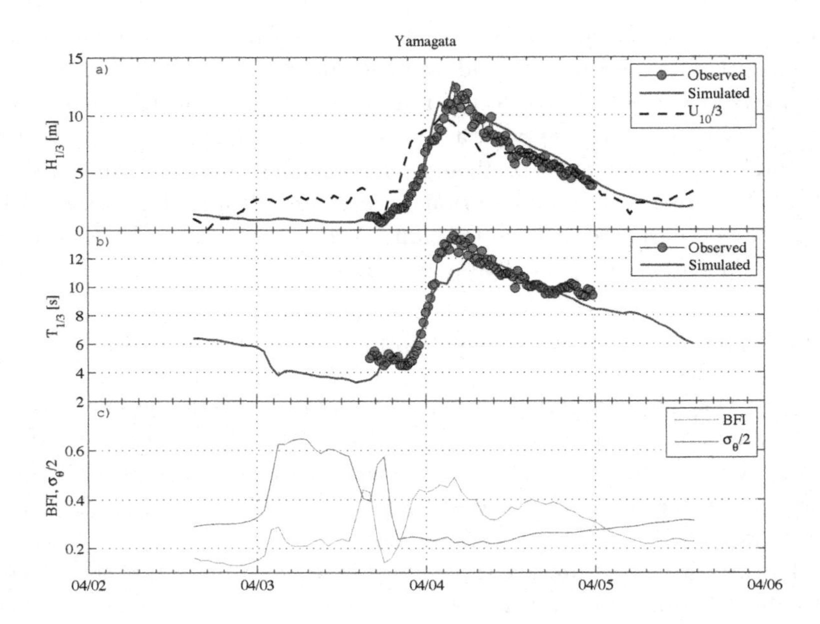

Fig. 9.7. Example of a severe winter storm in the Sea of Japan in April 2012 (upper: significant wave height, middle: significant wave period, bottom: directional spread; lines: observed data, marks: hindcast).

consists of random waves as a series of weakly correlated sequential waves as shown in Fig. 9.8. The measured offshore wave profile is sometimes used as an offshore boundary condition for a three dimensional model or Boussinesq type model in the nearshore. However, an offshore wave condition widely used is a JONSWAP-like parameterized spectra of linear random waves for a given significant wave height. Therefore, a mismatch still exists between offshore and nearshore modeling in engineering applications.

One major cause of coastal disasters is unexpectedly large waves due to storms or tropical cyclones. Researchers expect an increase in intensified tropical cyclones due to global climate change at the end of the 21st century.[8] Greenhouse effects on global climate change will influence coastal areas and require impact assessments and adaptation strategies for the future. Sea level rise is a static side issue of climate change, and it is an important consideration for human activity near the coastal zone. From 1870 to 2004, the global sea level rose by 1.7 ± 0.3 mm/year.[8] On the other hand, future changes in ocean wave climates are a dynamic side is-

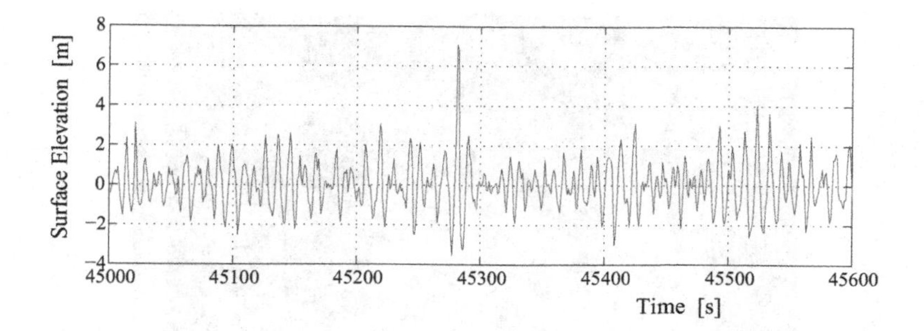

Fig. 9.8. Example of observed surface elevations offshore the Sea of Japan.

sue of how climate change influences coastal regions.[11] If extreme weather events become stronger in the future, it is necessary to seriously consider the effects of these dynamic phenomena to prevent and reduce the impact of coastal disasters. Understanding how storm waves will change in the future is an important consideration for safety both near and offshore of the coast. For example, a coastal breakwater is designed with a maximum storm surge level and a maximum dynamic pressure from a maximum wave condition for a predetermined design lifetime. To assess climate change impacts, it is important to know how wave loads will vary with designed extreme wave conditions as the climate warms. It is important to know climate influences on ocean waves and sea level rise, and how those influences will combine before a severe condition occurs. Numerical experiments using three dimensional models are quite useful to assess impacts and adaptation strategies.

With climate change, chemical and biological ocean environments are also anticipated to change via modifications of heat and gas exchanges between the atmosphere and the ocean. While global assessment of the flux transfer across the sea surface has been computationally undertaken by a general circulation model (GCM) coupled with an ocean current model, for rational, regional predictions of the heat and moisture budgets, the underlying responses of local surface processes for climate change should be understood. In particular, the main causes that enhance material transfer from the atmosphere to the ocean are known to be turbulence at the sea surface due to wind and wave breaking as well as aeration induced by surface turbulence. Turbulence disturbs the skin of the sea surface, where gas is highly dissolved, and transports air downward,[15] creating a so-called surface-renewal process,[10] and allowing the air-bubbles under the surface

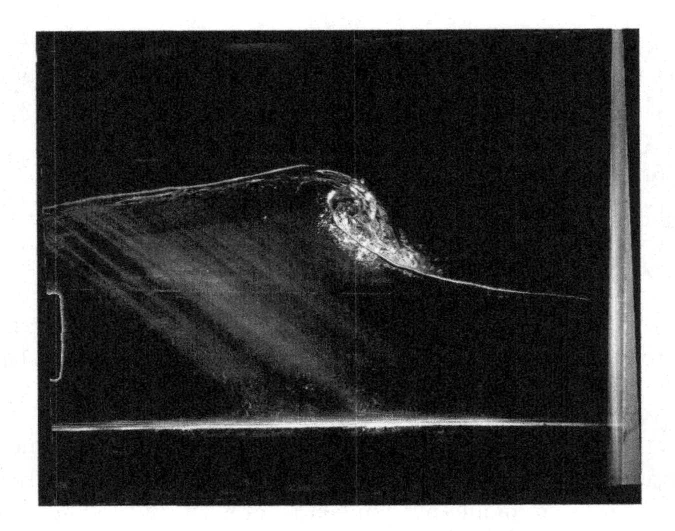

Fig. 9.9. Snapshot image of a breaking wave in the surf zone.[12]

to transfer gas into the depths.[3] As introduced in §5.5, gas-liquid flow computations with wind-wave interactions have already been undertaken, and therefore further improvements of the NWF model may contribute to computational assessments of this type of interaction in the future.

Wave breaking creates dense plumes of bubbles, dissipating energy and momentum. Peregrine[14] summarized that entrapped air in breaking waves reduces the wave impact pressure since the air has greater compressibility compared with a single-phase approach. The entrapped air reduces the pressure approximately 10%. The enhanced compressibility of the air-water mixture decreases the velocity of sound and is being used to estimate large-scale prototype impacts, since the usual Froude scaling is unlikely to be correct for engineering problems. Therefore, the connection between air-water mixture, bubble distribution, and wave breaking induced turbulence is essential to understand gas-liquid interactions in breaking waves in the surf zone. An accurate estimate of bubble size and population distributions in the surf zone is important for understanding two-phase flow characteristics and solving engineering problems and environmental mechanisms in the coastal area. Recent imaging studies have illustrated the disintegration of entrapped air cavities dividing into bubbles.[1,2] Figure 9.10 shows the initial stage of bubble entrainment in a breaking wave in the surf zone, and such microscopic scales create several complex phenonema as injection of

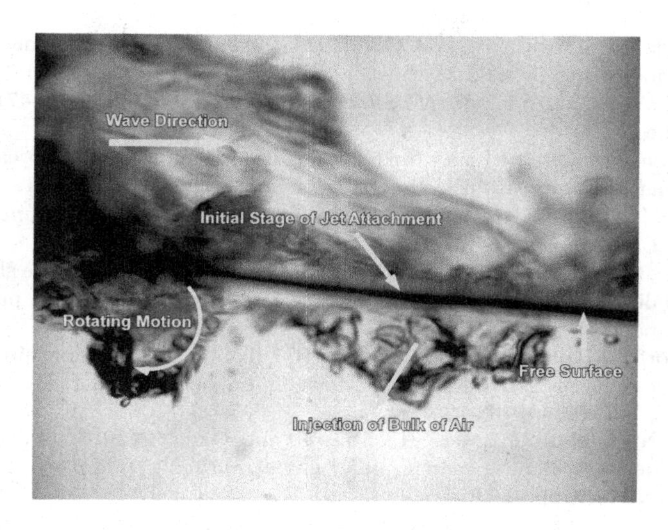

Fig. 9.10. Bubble injection due to a breaking wave.[12]

bulk air, bubble generation, etc. Two phase modeling of ocean waves is not yet well established without vague empirical element models. There are unexplained aspects of the problem, such as scale effects of voids and bubble size distributions in laboratory experiments, and the relation between void fraction, turbulence, and enhanced bubble populations in saltwater. Specifically, the processes of bubble injection and coalescence on subgrid size scales are important topics for numerical modeling in the future.

References

1. Deane G.B. and Stokes M.D. (1999): Air entrainment processes and bubble size distributions in the surf zone, *Journal of Physical Oceanography*, Vol.29, pp.1393-1403.
2. Deane G.B. and Stokes M.D. (2002): Scale dependence of bubble creation mechanisms in breaking waves, *Nature*, Vol.418, pp.839-844.
3. Farmer D.M., McNeil C.L. and Johnson B.D. (1993), Evidence for the importance of bubbles in increasing air-sea gas flux, *Nature*, Vol.361, pp.620-623.
4. Goda Y. (2000): *Random seas and design of maritime structures*, World Scientific.
5. Goda Y. and Takagi H. (2000): A reliability design method of caisson breakwaters with optimal wave heights, *Coastal Engineering Journal*, Vol.42, pp.357-387.

6. Harada E., Tsuruta N. and Gotoh H. (2011): Large eddy simulation for vertical sorting process in sheet-flow regime, *Journal of Japan Society of Civil Engineers, Ser. B2 (Coastal Engineering)*, Vol.67(2), pp.471-475 (in Japanese).
7. Hudson R.Y. (1959): Laboratory investigation of rubble mound breakwaters, *Journal of Waterways and Harbor Div.*, Vol.85, WW3, pp.93-121.
8. IPCC (2007): *Fourth Assessment Report, Intergovernmental Panel on Climate Change.*
9. Ito K., Oda Y. and Toe T. (2002): Study on deformation of gentle slope-type sea wall, *Annual Journal of civil engineering in the ocean*, Vol.18, pp.245-250 (in Japanese).
10. Komori S., Murakami Y. and Ueda H. (1989): The relationship between surface-renewal and bursting motions in an open-channel flow, *J. Fluid Mech.*, Vol.203, pp.103-123.
11. Mori N., Yasuda T., Mase H., Tom T. and Oku Y. (2010): Projection of extreme wave climate change under global warming, *Hydrological Research Letters*, Vol.4, pp.15-19.
12. Mori N., Kakuno S. and Cox D.T. (2009): Aeration and bubbles in the surf zone, in *Hand book of Coastal and Ocean Engineering*, Ed. Young C. Kim, Chapter 5, World Scientific Pub. Co., pp.115-130.
13. Morison J.R., O'Brien M.P., Johnson J.W. and Schaaf S.A. (1950): The force exerted by surface waves on piles, *Petroleum Trans.*, Vol.189, TP 2846, pp.149-154.
14. Peregrine D.H. (2003): Water wave impact on walls, *Annual Review of Fluid Mechanics*, Vol.35, pp.23-43.
15. Watanabe Y. and Mori N. (2008): Infrared measurements of surface renewal and subsurface vortices in nearshore breaking waves. *J. Geophys. Res.*, Vol.113, C07015, doi:10.1029/2006JC003950.

Acronyms List

Chapter 2

DNS Direct Numerical Simulation
GLS Generic Length Scale
GS Grid Scale
LES Large Eddy Simulation
NWF Numerical Wave Flume
NWT Numerical Wave Tank
RANS Reynolds Averaged Navier-Stokes
SG Subgrid
URANS Unsteady Reynolds Averaged Numerical Simulations

Chapter 3

BiCG Bi-Conjugate Gradient
BiCGStab Bi-Conjugate Gradient Stabilized
ADI Alternating Direction Implicit
CG Conjugate Gradient
FTCS Forward-Time Central-Space
GMRES Generalized Minimal RESidual
ICCG Incomplete Cholesky Conjugate Gradient
MICCG Modified Incomplete Cholesky Conjugate
PCG Preconditioned Conjugate Gradient
SOR Successive Over-Relaxation
TBL Turbulent Boundary Layer
VOF Volume of Fluid

Chapter 4

CADMAS-SURF SUper Roller Flume for Computer Aided Design of MAritime Structures
CFL Courant-Friedrichs-Lewy
MILU-BiCGSTAB Modified Incomplete Lower-Upper factorization-BiConjugate Gradient STABilized
NASA National Aeronautics and Space Administration
SMAC Simplified Marker And Cell
QUICK Quadratic Upstream Interpolation Convective Kinematics

Chapter 5

CCUP CIP Combined, Unified Procedure
CIP Constrained Interpolation Profile
CIPCLS CIP Conservative semi-Lagrangian Scheme
CSF Continuum Surface Force

Chapter 6

CISPH Corrected Incompressible Smooth Particle Hydrodynamics
CMPS Corrected Moving Particle Semi-implicit
CMPS-HMS Corrected Moving Particle Semi-implicit with Higher order Modified Source terms
CMPS-HS Corrected Moving Particle Semi-implicit with Higher order Source terms
CISPH-HS Corrected Incompressible Smooth Particle Hydrodynamics with Higher order Source terms
CMPS-SBV Corrected MPS with Strain-Based Viscosity
CSPH Corrected Smooth Particle Hydrodynamics
DEM Distinct Element Method
DSMC Direct Simulation Monte Carlo
-ECS Error Compensating Source
-GC Gradient Correction
-HL Higher order Laplacian
-HS Higher order Source
ISPH Incompressible Smooth Particle Hydrodynamics
LBM Lattice Boltzmann Method
LES Large Eddy Simulation

MPS Moving Particle Semi-implicit
MPS-WC Moving Particle Semi-implicit with Weak Compressibility
PPE Poisson Pressure Equation
PS Particle Scale
SGS Sub-Grid Scale
SPH Smoothed Particle Hydrodynamics
SPS Sub-Particle-Scale
WCSPH Weakly Compressible Smooth Particle Hydrodynamics

Chapter 7

DEM Distinct Element Method

Chapter 8

MAC Maker-and-Cell
PIC Particle in Cell

Chapter 9

CWD Computational Wave Dynamics
CFD Computational Fluid Dynamics
GCM General Circulation Model
NWF Numerical Wave Flume
NWT Numerical Wave Tank

Index

M

N

O

P

Q

R

S